COMPUTING

A Historical and Technical Perspective

COMPUTING
A Historical and Technical Perspective

Yoshihide Igarashi
Gunma University
Kiryu, Japan

Tom Altman
University of Colorado Denver
USA

Mariko Funada
Hakuoh University
Oyama, Japan

Barbara Kamiyama
Gunma University
Kiryu, Japan

CRC Press
Taylor & Francis Group
Boca Raton London New York

CRC Press is an imprint of the
Taylor & Francis Group, an **informa** business

A CHAPMAN & HALL BOOK

CRC Press
Taylor & Francis Group
6000 Broken Sound Parkway NW, Suite 300
Boca Raton, FL 33487-2742

Printed on acid-free paper
Version Date: 20140227

International Standard Book Number-13: 978-1-4822-2741-3 (Paperback)

Library of Congress Cataloging-in-Publication Data

Igarashi, Yoshihide.
    Computing : a historical and technical perspective / Yoshihide Igarashi, Tom Altman, Mariko Funada, and Barbara Kamiyama.
        pages cm
    Includes bibliographical references and index.
    ISBN 978-1-4822-2741-3 (alk. paper)
    1. Computer science. I. Altman, Tom. II. Funada, Mariko. III. Kamiyama, Barbara. IV. Title.

QA76.I23 2014
004--dc23

2013049662

Visit the Taylor & Francis Web site at
http://www.taylorandfrancis.com

and the CRC Press Web site at
http://www.crcpress.com

Contents

Preface, xiii

Acknowledgments, xv

About the Authors, xvii

CHAPTER 1: THE DAWN OF COUNTING   1

1.1   ARCHEOLOGICAL EVIDENCE: PALEOLITHIC ART   1

1.2   FINGERS FOR COUNTING   2

1.3   THE USE OF TALLY STICKS AND REPRESENTATIONAL SYMBOLS: THE FIRST INFORMATION REVOLUTION   2

1.4   COUNTING BY PEBBLES   4

1.5   THE USE OF TOKENS AND THE SECOND INFORMATION REVOLUTION   5

REFERENCES   6

CHAPTER 2: REPRESENTATION OF NUMBERS   7

2.1   POSITIONAL NUMBER SYSTEMS   8

2.2   MORE ABOUT NUMBER SYSTEMS   10

2.3   FURTHER DISCUSSIONS OF ZERO   10

REFERENCES   14

CHAPTER 3: RATIONAL AND IRRATIONAL NUMBERS   15

3.1   APPEARANCE OF FRACTIONS   15

3.2   RATIONAL NUMBERS   17

3.3   IRRATIONAL NUMBERS   19

REFERENCES   21

CHAPTER 4: PRIME NUMBERS   23

   4.1   THE STORY OF PRIME   23

   4.2   THE PRIME NUMBER THEOREM   29

   REFERENCES   33

CHAPTER 5: EUCLID'S ELEMENTS   35

   REFERENCES   40

CHAPTER 6: DIOPHANTUS OF ALEXANDRIA AND *ARITHMETICA*   43

   REFERENCES   49

CHAPTER 7: SECRET WRITING IN ANCIENT CIVILIZATION   51

   7.1   STEGANOGRAPHY   51

   7.2   CRYPTOGRAPHY   52

   REFERENCES   57

CHAPTER 8: THE ABACUS   59

   8.1   THE EARLIEST ABACI   59

   8.2   THE SALAMIS TABLET AND THE ROMAN
          HAND ABACUS   61

   8.3   THE CHINESE ABACUS   64

   8.4   THE JAPANESE ABACUS   65

   REFERENCES   66

CHAPTER 9: BOOK OF CALCULATION BY FIBONACCI   69

   REFERENCES   75

CHAPTER 10: DECIMAL FRACTIONS AND LOGARITHMS   77

   10.1 APPEARANCE OF DECIMAL FRACTIONS   77

   10.2 LOGARITHMS   79

   REFERENCES   83

CHAPTER 11: CALCULATING MACHINES   85

   11.1 THE *RECHEN UHR* OR "CALCULATING CLOCK"
         OF WILHELM SCHICKARD   86

11.2 THE PASCALINE   87

11.3 LEIBNIZ AND THE STEPPED RECKONER   88

11.4 THE JACQUARD LOOM   89

11.5 BABBAGE'S MECHANICAL COMPUTERS   91

11.6 ADA LOVELACE, THE FIRST COMPUTER PROGRAMMER   92

11.7 HERMAN HOLLERITH AND HIS AMAZING TABULATOR   93

REFERENCES   96

CHAPTER 12: SOLUTIONS TO ALGEBRAIC EQUATIONS   97

12.1 LINEAR EQUATIONS   98

12.2 QUADRATIC EQUATIONS   99

12.3 CUBIC EQUATIONS   100

12.4 QUARTIC AND QUINTIC EQUATIONS   101

REFERENCES   105

CHAPTER 13: REAL AND COMPLEX NUMBERS   107

13.1 REAL NUMBERS   107

13.2 COMPLEX NUMBERS   110

13.3 COMPLEX-VALUED FUNCTIONS   112

REFERENCES   113

CHAPTER 14: CARDINALITY   115

REFERENCES   120

CHAPTER 15: BOOLEAN ALGEBRAS AND APPLICATIONS   121

REFERENCES   128

CHAPTER 16: COMPUTABILITY AND ITS LIMITATIONS   129

16.1 GÖDEL'S INCOMPLETENESS THEOREM   129

16.2 TOTAL FUNCTIONS   130

16.3 TURING MACHINES   131

16.4 CHURCH–TURING'S THESIS   134

REFERENCES   136

CHAPTER 17: CRYPTOGRAPHY FROM THE MEDIEVAL TO THE MODERN AGES 137

17.1 THE ARAB CRYPTANALYSTS   137
17.2 POLYALPHABETIC SUBSTITUTION CIPHERS   139
17.3 HOMOPHONIC SUBSTITUTION CIPHERS   141
17.4 ENIGMA MACHINE   143
17.5 BREAKING ENIGMA CODES   144
17.6 LORENZ CIPHER   145
REFERENCES   146

CHAPTER 18: ELECTRONIC COMPUTERS   149

18.1 THE ABC COMPUTER   149
18.2 THE Z3 COMPUTER   150
18.3 THE COLOSSUS COMPUTER   151
18.4 THE ENIAC COMPUTER   153
18.5 VON NEUMANN ARCHITECTURE FOR COMPUTERS   155
18.6 OTHER NOTABLE EARLY ELECTRONIC COMPUTERS   156
    18.6.1 National Physics Laboratory and the ACE   156
    18.6.2 The MARK 1 at Manchester University   157
    18.6.3 Electronic Delay Storage Automatic Calculator
            (EDSAC)   158
    18.6.4 Whirlwind I   158
    18.6.5 Standards Eastern Automatic Computer (SEAC) 158
    18.6.6 Standards Western Automatic Computer (SWAC) 158
REFERENCES   158

CHAPTER 19: NUMERICAL METHODS   161

19.1 NUMERICAL CALCULATION IN ANCIENT
     CIVILIZATIONS   161
19.2 NUMERICAL SOLUTION OF ALGEBRAIC EQUATIONS   164
19.3 MODERN NUMERICAL ANALYSIS AND ITS
     PROBLEM DOMAINS   170
REFERENCES   172

CHAPTER 20: MODULAR ARITHMETIC | 173

20.1 CLOCK ARITHMETIC | 173

20.2 CHINESE REMAINDER THEOREM | 175

20.3 FERMAT'S LITTLE THEOREM | 178

REFERENCES | 179

CHAPTER 21: CYBERNETICS AND INFORMATION THEORY | 181

21.1 NORBERT WIENER AND CYBERNETICS | 181

21.2 SHANNON'S INFORMATION THEORY | 183

21.3 SHANNON–FANO CODING AND HUFFMAN CODING | 186

21.4 MORSE CODE | 189

REFERENCES | 190

CHAPTER 22: ERROR DETECTING AND CORRECTING CODES | 193

22.1 PARITY CHECK CODES | 193

22.2 HAMMING CODES | 194

22.3 LINEAR CODES | 198

REFERENCES | 202

CHAPTER 23: AUTOMATA AND FORMAL LANGUAGES | 205

23.1 AUTONOMOUS APPARATUS | 205

23.2 AUTOMATA AS COMPUTING MODELS | 206

23.3 FORMAL LANGUAGES | 211

REFERENCES | 214

CHAPTER 24: ARTIFICIAL INTELLIGENCE | 217

24.1 WHAT IS AI? | 218

24.2 AI TIMELINE | 219

24.3 AI PIONEERS | 224

24.4 AREAS OF AI | 227

REFERENCES | 229

Chapter 25: Programming Languages   231

25.1   MACHINE CODE   231

25.2   INTERPRETATIVE CRUTCHES   232

25.3   THE FIRST HIGH-LEVEL LANGUAGE: FORTRAN   232

25.4   OVERVIEW: IMPERATIVE PROGRAMMING   233

25.5   OVERVIEW: DECLARATIVE PROGRAMMING   234

25.6   THE SECOND HIGH-LEVEL LANGUAGE: LISP   234

25.7   OVERVIEW: FUNCTIONAL PROGRAMMING   235

25.8   STANDARDIZATION AND COMPROMISE:
        ALGOL 60   235

25.9   FROM SCIENCE TO BUSINESS: COBOL   237

25.10   BACK TO THE BASICS   238

25.11   OVERVIEW: LOGICAL PROGRAMMING   239

25.12   PROGRAMMING LOGIC: PROLOG   239

25.13   OVERVIEW: OBJECT-ORIENTED PROGRAMMING   239

25.14   THE FIRST OBJECT-ORIENTED PROGRAMMING
        LANGUAGE: SMALLTALK   240

25.15   IMPERATIVE AND OBJECT ORIENTED: C++   240

25.16   OBJECT ORIENTED, HOLD THE IMPERATIVE: JAVA   241

25.17   THE BEST OF BOTH WORLDS: C#   242

REFERENCES   244

Chapter 26: Algorithms and Computational Complexity   245

REFERENCES   253

Chapter 27: The Design of Computer Algorithms   255

27.1   SORTING AND SEARCHING   255

27.2   DATA STRUCTURES   258

27.3   GRAPH ALGORITHMS   260

27.4   RANDOMIZED ALGORITHMS   265

REFERENCES   268

CHAPTER 28: PARALLEL AND DISTRIBUTED COMPUTING   271

   28.1 DAWN OF PARALLELISM   271

   28.2 PARALLEL COMPUTERS   273

   28.3 PARALLEL ALGORITHMS   275

   28.4 DISTRIBUTED COMPUTING   279

   REFERENCES   281

CHAPTER 29: COMPUTER NETWORKS   283

   29.1 PACKET SWITCHING NETWORKS   283

   29.2 ARPANET AND CSNET   284

   29.3 WORLD WIDE WEB   287

   29.4 CLOUD AND GRID COMPUTING   289

   29.5 UBIQUITOUS COMPUTING   291

   REFERENCES   292

CHAPTER 30: PUBLIC-KEY CRYPTOGRAPHY   295

   30.1 THE SITUATION IN THE 1960s AND 1970s
       BEFORE THE PUBLIC KEYS   295

   30.2 THE BIRTH OF PUBLIC-KEY CRYPTOGRAPHY   297

   30.3 RSA CRYPTOGRAPHY   300

   30.4 DIGITAL SIGNATURES   303

   30.5 ANOTHER STORY OF PUBLIC-KEY CRYPTOGRAPHY
       FROM ENGLAND   305

   REFERENCES   306

CHAPTER 31: QUANTUM COMPUTING   309

   31.1 THE BASICS OF QUANTUM COMPUTING   309

   31.2 QUANTUM COMPUTATION LOGIC AND GATES   312

   31.3 FAMOUS QUANTUM ALGORITHMS   312

      31.3.1 Deutsch's Algorithm (1989)   313

      31.3.2 Grover's Search Algorithm (1995)   313

      31.3.3 Shor's Factoring Algorithm (1994)   314

31.4 DIFFICULTIES AND LIMITS OF
QUANTUM COMPUTING   315

31.5 CLOSING SUMMARY   316

REFERENCES   316

Index   319

Preface

Have you ever wondered why the way we count differs greatly from the way modern-day electronic computers do? Or why there are 24 hours in a day, 60 minutes in an hour, etc.? How about who invented numbers and when they were invented, as well as why there are different kinds of numbers. Did you know that secret writings and cryptography date back to ancient civilizations? You will find the answers to these questions, and many more, in just the first few chapters of this book.

The primary purpose of this book is to serve as a supplementary textbook for a mid- to upper-level university undergraduate course offered to science and engineering majors. The book contents are self-explanatory, and no prior knowledge of advanced mathematics or computer science is needed. While some chapters require much deeper reading and analysis than others, the book's primary purpose is to expose the reader to the vast array of topics related to computation. This includes purely historical as well as the technical aspects of ancient and modern-day computing. The organization of the 31 chapters follows the historical timeline. Therefore, while from chapter to chapter the topics may appear to be quite unrelated, each chapter is "stand-alone" and does not have to be read in sequence. The book's primary value is as a historical reference source in the field, together with brief technical explanations and examples.

The first 10 chapters address what might be considered historical accounts of counting and computing. It might surprise the reader to find that the ancient Greeks knew of the existence of numbers that cannot be measured. Also, it turns out that the modern-day numbers we call *Arabic numerals* actually have their origins in India. These and a number of historically interesting facts are discussed and explained, including examples from the original works of such renowned scholars as Pythagoras, Diophantus, Fibonacci, and many others.

Subsequent chapters include the early computing devices, such as the *abacus*, Schickard's calculating clock, the Pascaline, Babbage's mechanical computers (the difference engine and the analytical engine), etc. Various parallel developments in mathematics are listed as well. These include solutions to algebraic equations, real and complex numbers, and Cantor's unexpected result addressing the cardinalities of sets and the fact that there is more than one "infinity," and not only that, but also the number of these infinities itself is infinite.

The last portion of the book tends to focus on the notion of *computability*, its capabilities as well as its limitations. It is quite interesting from a historical perspective that the formal model of computation was introduced by Turing some 15 years before the first electronic computers were built. An even more surprising fact is that there must exist problems for which no computer solutions exist. While Turing's *halting problem* was the first of such problems, it turns out the set of problems that cannot be computed far outnumbers that which can (keeping in mind that both sets are infinite).

Additional important computational topics such as numerical methods, information and coding theory, automata and formal languages, and artificial intelligence are but a few of the computer science–related chapters that follow. Also included are some practical issues within this area, such as computational complexity, parallel computation, computer networks, and public-key cryptography. The last chapter of the book is a brief introduction to quantum computation, an area of computer science still in its infancy, but showing a great deal of potential.

While the reader can find a great deal of information about the book's topics on the Internet, we hope that by organizing it here, a historical perspective will allow the reader not only to gain insight into each individual topic, but also to develop a deep appreciation for the long evolutionary processes over the millennia. During many cultures, innumerable individuals have contributed their talents and creativity to formulate what has become our mathematical and computing heritage. Since each of the chapters has been written more or less independently, it is noteworthy that we have actually learned a significant amount while researching our assigned topics, and especially when proofreading each other's chapters.

Acknowledgments

This book would not have been possible without the support and encouragement of many people, all of whom are our friends, colleagues, and students. Our deepest thanks go to Maarten H. van Emden, Akira Maruoka, Forbes D. Lewis, Geoff Dowling, Robert Senser, and Christine Coder.

We also express our appreciation to Randi Cohen, Marsha Pronin, and Judith Simon of CRC Press for making our job so much smoother and easier.

Special thanks go to our families, Sachiko, Hania, Hiroshi, Yoshiko, Robert, Tadashi, and Johannah, for their patience, understanding, and having to put up with us over the past 5 years.

About the Authors

Yoshihide Igarashi is a professor emeritus of computer science at Gunma University, Kiryu, Japan. He received his Ph.D. from Tohoku University, Sendai, Japan (1971). He worked in the School of Artificial Intelligence, the University of Edinburgh (1972–1974). He taught theoretical computer science at the University of Leeds (1974–1977), at City University, London (1977–1978), Gunma University (1978–2004), and at the University of Kentucky (1980–1981 and 1987–1988). His research interests include computational complexity, the design and analysis of algorithms, and parallel and distributed computing. He is an author of over 150 technical, conference, and journal papers.

Tom Altman is a professor of computer science at the University of Colorado, Denver. He holds a Ph.D. from the University of Pittsburgh (1984). Currently, he is teaching automata theory, theory of computation, algorithms, and computational complexity. His research interests include optimization algorithms, parallel computation, and bioinformatics. He has authored over 80 research papers and is a recipient of several teaching and research awards in his field.

Mariko Funada is a professor of business administration at Hakuoh University, Oyama, Japan. She received her Ph.D. from Tokyo Institute of Technology, Tokyo, Japan, in 1992. She has been teaching computer science and information processing at Hakuoh University and Gunma University. Her research interests include human-computer interaction (HCI) and multivariate data analysis. She is an author of over 60 technical, conference, and journal papers.

Barbara Kamiyama has been a lecturer of technical English at Gunma University, Kiryu, Japan since 1994. She received a B.A. in social science from Fordham University, New York (1972), and attended graduate school in the Department of Anthropology at the State University of New York at Binghamton (1974–1977) and in the Department of Computer Science at Vanderbilt University, Nashville, Tennessee (1978–1979). Her research interests include the archeology and anthropology of mathematics and computing.

CHAPTER 1

The Dawn of Counting

The origin of abstract numbers is lost forever in the mists of time. To know what lies behind that thick curtain of mist we call human history, we must sift through what has been excavated archeologically or decipher clues in what remains in the written record of civilizations, also long gone. In doing so, we may catch a glimpse of what first motivated us to invent numbers. It is important to remember that the invention of numbers was not a spontaneous act, but rather the result of a long evolutionary process.

Animal behaviorists have observed that birds have some awareness of the size of a collection of objects in terms of the number of eggs in a nest and can detect changes in that number. Wasps have been observed to feed a set number of caterpillars to male eggs and twice as many to female eggs. It is likely that, in the case of human evolution, a similar ability to discriminate variation in the size of a collection is what eventually, through trial and error, led to the ability to conceive of abstract numbers. Let's look at what evidence we have, sketchy though it may be.

1.1 ARCHEOLOGICAL EVIDENCE: PALEOLITHIC ART

The dramatic discovery of cave art in the south of France has changed forever our understanding of the *Homo sapiens* we know as Cro-Magnon and who lived there during the Paleolithic era, or about 30,000 years ago. As spelunkers cast their lantern light on the walls of caves left hermetically sealed for thousands of years, wild horses, buffalo, and other extinct animals thundered across the stone face of the walls in a breathtaking panorama! Here was a clear effort by gifted Paleolithic artists to represent the large herds of wild animals with which these communities shared their environment, among which they foraged for food, and that inspired them with both fear and reverence. Surreal red handprints on the cave walls reach out from the past into the present, while collections of red-ochre dots may represent some sort of calendar—perhaps counting the phases of the moon. In one cave, a mural depicts hunters posing with bows and arrows, lined up in a neat one-to-one correspondence with deer as their prey. All of these paintings tell the tale of a human awareness of numbers—not necessarily the abstract concept of numbers or words that represent our understanding of numbers—but surely the awareness of the size of a collection, the ability to represent a succession of events as in the use of a simple calendar, and the ability to make a one-to-one correspondence between different sets. We were well on our way to counting!

It is highly unlikely that Paleolithic humans had numbers as we know them today. One possible explanation is that the invention of abstract numbers is a product of social necessity. Since the earliest humans probably roamed seasonally over their environment and had few possessions, there was little need for creating specific words for large, abstract numbers; sets or groups of objects (e.g., herds of animals, fish in the river, etc.) were simply referred to as *very many* in some primitive cultures, and the ability to count was limited to three words: *one*, *two*, and *many*.

1.2 FINGERS FOR COUNTING

Although there is no definitive proof, many evolutionists think that the first tools humans used for counting were their fingers. With a naïve ability to make a one-to-one correspondence, the use of fingers seems almost inevitable. The word *digit* in English indicates how natural this tendency is among humans. Digit derives from the Latin word *digitus*, which in that language means "finger." In English, however, the word *digit* not only means finger, but also a discrete number.

Various ancient civilizations devised clever methods for expanding the count by using finger joints, whole fingers, or including other parts of the body as well. While using fingers (or other body parts) can be very convenient, this method has one serious drawback: fingers leave no record of what is counted unless some other tool is added to the technique. Inevitably, humans, compelled by need, sought better methods.

1.3 THE USE OF TALLY STICKS AND REPRESENTATIONAL SYMBOLS: THE FIRST INFORMATION REVOLUTION

Carving notches on bones or stones has been called by some the first information revolution. Alexander Marshack (1918–2004), in his seminal work

The Roots of Civilizations (1972) [2], demonstrated that Upper Paleolithic carvings on bones, dating back some 32,000 years, represented lunar calendars. The ability to make meaningful observations and record them on a surface constitutes a significant step in human evolution and represents man's first attempt at data storage and retrieval. While many similar artifacts have been excavated from Cro-Magnon settlements, Neanderthals living during the same period did not seem to be able to employ representational symbols. It has been suggested by some that this may explain in part why Neanderthals eventually died out.

Other important Paleolithic evidence of techniques for counting comes to us from South Africa. A long piece of bone, known as the Lebombo stick, named after the cave in which it was found, dates back about 35,000 years. This bone has 29 tally marks on it, indicating that it was used as a means of keeping a count. Whether it was objects being counted or was used as a sort of calendar, no one knows. Here again we see the ability to make a one-to-one correspondence between the objects or events being counted and the marks on the bone. Each notch represents one object. From such evidence, however, we can draw the conclusion that humans had the ability and desire to keep a running tally of something from the very dawn of the species.

The Ishango bone from the head of the Nile River in the East Africa area dates from around 20,000 years ago. It is an interesting artifact because its notches are arranged in groups, which may indicate an attempt to further facilitate counting by creating a higher level of multiples. In other words, there may have been one word in the language for a single unit and another word representing grouped units. All of this is conjecture, but clearly this was another possible step in the evolution of numbers as we know them today.

Tally sticks have been widely used not only in Africa but also throughout Europe. Paper, as invented by the Chinese, was not introduced in Europe until the 14th century. Until then, manuscripts were written on parchment. Even with the introduction of paper, however, few could read or write, making the continued use of tally sticks an important method for ordinary people to keep track of transactions, whether buying bread in a bakery, paying taxes, or recording the amount of milk produced by a herd of cows.

A later adjustment was the introduction of the split tally stick, which, once notched, could be divided lengthwise to allow both parties in a transaction to hold a record of it. Eventually, the longer portion of the stick became known as the stock and was kept by the person who had lent money or sold goods (and from which we get the term *stockholder*), and the shorter portion, given to the customer, was known as the foil. It is from this custom that we get the expression "the short end of the stick," indicating the less admirable end of a deal.

Tally marks and tally sticks remain in use to the present day. Keeping a tally is a remarkably simple way of performing addition without using numbers at all. By carving notches into the bone or stick, it becomes, at the same time, a lasting record of the items being counted.

1.4 COUNTING BY PEBBLES

In addition to tally sticks, early humans—again without the use of abstract numbers—probably used pebbles to count the number of objects in a collection by way of one-to-one correspondence. One advantage of this system over the use of fingers is that, besides being able to count past 10, it does leave a physical record of the count.

As the size of human settlements increased, and people became more sedentary, it was also possible to own more objects. This provided the impetus to invent better systems for counting. Many researchers have given the example of a shepherd counting his sheep as they are led out to pasture by dropping a pebble for each animal as it leaves the corral. Upon their return at some later point, the shepherd uses the pebbles from the previous count to do a recount. If more sheep return than he has pebbles, he knows his flock has somehow increased, and conversely, if fewer return than the number of pebbles used when they departed, he knows he has lost some sheep and must go in search of them. This is a very simple, but accurate, means of keeping a count, requiring neither concrete nor abstract numbers. Such a method can be applied to bags of grain, jars of oil, and other objects used in everyday life.

As quantities being counted grew in size, pebbles could be differentiated such that varying colors or sizes might represent multiples, thus obviating the need for carrying about increasingly heavy bags of pebbles. This is similar to the practice of making groups of notches on a tally stick to represent a multiple. The principle of one-to-one correspondence still applies since one type of pebble represents one size of a group of objects. The use of both tally sticks and pebbles, precursors of numbers, was an important step in the evolution of counting that eventually led to the formation of abstract numbers.

The word *calculate* in English, which means "to count," indicates the importance of the use of pebbles as a means of counting. The word is derived from the Latin word *calculus*, which means "pebble" or "stone." The same Latin word is also the root of the word *calcium* in English, which is a mineral. In other words, in English, the Latin root word *calculus* is used to form words meaning both count and a kind of mineral. Here again, like the word *digit*, the use of the Latin root word in English demonstrates the close association of two completely different concepts as a result of a common practice such as using pebbles to count objects.

1.5 THE USE OF TOKENS AND THE SECOND INFORMATION REVOLUTION

It took several millennia for humans to move from the use of fingers to tally sticks and pebbles. Such practices remain in human societies even today. With the development of agriculture and the establishment of cities by the first civilizations, however, these common practices, critical steps in the evolution of numbers, finally gave way to more sophisticated techniques. The ability to plant and raise cereal crops allowed for rapid population growth. Archeological excavations in the Middle East, where the earliest cites were established, have discovered large numbers of small clay objects of various shapes and sizes that served as tokens for counting. Different shapes represented different types of goods.

As products were traded at increasingly long distances, clay tokens were used as bills of lading. Eventually loose clay tokens were pressed into clay tablets, to make an impression. According to archeologist Denise Schmandt-Besserat (1933–) [4], this represents a movement from the use of three-dimensional objects to a two-dimensional record, and the impetus behind the second information revolution, the replacement of objects for symbols in a representational system.

Michael Rothschild (1942–) has written about the eventual adaptation of clay tablets in Sumerian civilization about 5,000 years ago and the appearance of the first writing system [1]. The vast majority of tablets excavated in Sumeria are documents related to accounting. With the rise of civilization and the need to collect taxes, maintain a bureaucracy, and have the ability to produce, import, and export increasingly larger quantities of goods came a simultaneous need to keep and maintain records and employ larger-and-larger-scale numbers. Humans had moved into the world of abstract numbers, and the archeological record suggests that it was the compelling need to keep accounts that stimulated the rise of the first counting and writing systems.

REFERENCES

1. M. Rothschild, Cro-Magnon's Secret Weapon, *Forbes ASAP: A Technology Supplement*, September 13, 1993.
2. A. Marshack, *The Roots of Civilizations: The Cognitive Beginnings of Man's First Art, Symbol and Notation*, McGraw Hill, New York, 1991.
3. J. MacLeish, *The Story of Numbers*, Fawcett Columbine, New York, 1991.
4. D. Schmandt-Besserat, *How Writing Came About*, University of Texas Press, Austin, 1996.
5. E. Ascalone, *Mesopotamia*, University of California Press, Berkeley, 2005.
6. H. Blohm, S. Beer, and D. Suzuki, *Pebbles to Computers: The Thread*, Oxford University Press, Toronto, 1986.

CHAPTER 2

Representation of Numbers

In simple societies, tally sticks and clay tokens served adequately for thousands of years to represent and record objects counted. As the number of these objects increased, however, various methods appeared to group the notches on sticks or bones or to designate tokens of higher orders in order to facilitate the process of counting and recording. With the advent of agriculture and the division of labor, increased population growth, and the formation of centralized bureaucracies, simple tools like tally sticks and clay tokens no longer sufficed.

It is believed that the earliest forms of written numbers evolved when the spoken names for numbers became associated with their recorded tallies. In addition, the process of going from a simple one-to-one correspondence type of tally to one in which tally marks or symbols are grouped is an important cognitive step that led eventually to the formation of number systems. Forming groups of various sizes and ordering them in progression led to the creation of the earliest number symbols and systems [8].

For example, in ancient Egypt, a simple one-stroke hieroglyphic tally recorded on papyrus evolved into a grouped tally, which then further evolved into a hieratic code system of abstract symbols. Similarly, the ancient Sumerians and Babylonians at first recorded wedge-shaped tallies on clay tablets. These cuneiform (from the Latin word *cuneus* for wedge) symbols were later grouped for recording speed and convenience, and then eventually made into graduated abstract numerals that formed the basis of a base-60 number system [7].

In most cultures it is clear that the words used to represent numbers evolved quite separately from their written representations. While the words for numbers evolved for the most part within societies and are directly linked to specific languages or language groups, the written Hindu-Arabic numerals we use today spread from country to country to become what Karl Menninger (1898–1963) referred to as "the most significant symbol of mankind's universality" [8].

2.1 POSITIONAL NUMBER SYSTEMS

According to the principle of *position*, also known as *place-value*, the value assigned to any given numeral depends on the position in which the numeral appears. For example, in our decimal number system, the numeral 5 appears three times in the number 555; however, each 5 has a different value because each 5 holds a different position. Reading from the right, the first 5 means 5 units of one, the next 5 to the left represents 5 tens or 50, and the next 5 to the left represents 5 hundreds or 500. We have become so accustomed to this written convention that it seems second nature; however, in actual fact, only a few civilizations ever invented this practice. Their invention proved so convenient and practical that positional notation spread throughout the world and is now the standard used everywhere.

The Babylonians were the first to invent a place-value number system. Evidence shows that as early as 2000 BC Babylonian scholars employed a written place-value base-60 or *sexagesimal* number system, carrying out quite advanced mathematical calculations for astronomical research and other forms of record keeping required by their bureaucracy. While their discovery of the principle of positional notation is the earliest known, their number system is not the one used in the modern world. Some cultural vestiges of their sexagesimal number system can be found in our convention of measuring 60 seconds per minute and 60 minutes per hour, and in calculations for angles and geographic longitude. The importance of this Babylonian invention can hardly be overstated, for without positional notation, different powers of a base required the use of different symbols, making calculation extremely difficult [1, 2].

The Chinese were the next to invent a positional number system around the second century BC. Unlike the Babylonians, theirs was a base-10 number system, with calculations performed on a precursor of the modern Chinese abacus, called *suan zi*, meaning "calculating with rods."

While there were written symbols for the numbers from 1 to 9 and for the powers of 10 (10, 100, 1000, and so on), these symbols were not used for calculation. Instead, the Chinese used bamboo or ivory rods on a checkerboard table. The columns on the table represented the powers of 10. This abacus allowed them to perform all arithmetic operations with rods, becoming a *de facto* positional notation system. Once calculations had been performed, the results were transcribed in written numerals, exactly paralleling the position of the rods on the table [1].

We come then, at last, to ancient India, which gave birth not only to the numerals used around the world today, but also to the positional number system and the concept of *zero* with which we are most familiar. The evolution of the nine numeral symbols is long and complicated. By the beginning of the fifth century BC, we find the units 1 to 9 represented by nine abstract Brahmi numerals in India [9]. At this time, Indian mathematicians employed a decimal place-value number system and had also invented the concept of zero. However, their zero was not a written numeral, but only a Sanskrit word.

The change to a decimal place-value number system with nine numerals came about as result of the use of a sand or dust abacus. The columns of this abacus corresponded to the powers of 10, and Brahmi numerals were used in the columns. Transcribing the results of calculation on the abacus into Sanskrit number words, Indian mathematicians mirrored the abacus format, arriving at a decimal place-value arrangement. After some time, mathematicians realized that the use of Brahmi numerals would be a much more efficient writing scheme and did away with writing out number words in Sanskrit. This evolution marks the birth of the numerals from 1 to 9 we use today.

The column format of a dust abacus allowed Indian mathematicians to calculate without a numeral for zero by just leaving the appropriate column blank. This became a problem, however, when transcribing calculations from the abacus to written notation. The Indians had a Sanskrit word for zero based on the concept of the sky or space. They took this word-symbol, usually either a circle or half-circle, and adapted it to represent their concept of zero. This came about sometime during the fourth century AD. It is at this point we find that the Indians used nine abstract numerals, employed a fully operational decimal place-value number system, and had invented a true zero. Although the numerals 0 to 9 are now commonly known as Arabic numerals, they are, as we have seen, Hindu-Arabic in origin [1].

2.2 MORE ABOUT NUMBER SYSTEMS

The decimal or base-10 number system was used in various ancient civilizations. It was most likely motivated by the common use of 10 fingers to count. Nondecimal systems, as we have seen in the previous section, have also occurred in various civilizations and in different eras. One example is the base-12 or duodecimal number system of the ancient Egyptians. Rather than counting fingers, they used finger joints or phalanxes for enumeration, a technique called *phalanx counting*. With the thumb as a pointer, each hand could be used to count up to 12, or up to a total of 24. The Egyptians used their duodecimal number system to divide a day into 24 periods, a custom that we continue to observe in our 24-hour day. Phalanx counting is still in use today in Egypt, Syria, Iraq, and some parts of South Asia.

The sexagesimal or base-60 system of the ancient Sumerians did not have unique symbols for the units from 1 to 59. Remembering that many unique symbols would have been burdensome, they subdivided the numbers into an auxiliary base of 10, with each group of 10 having a name. For example, one cuneiform represents 10, two cuneiforms represent 20, three cuneiforms represent 30, and so on.

Another example we might consider is Roman numerals. These numerals were used throughout the Roman Empire from ancient times. They dominated as the common numeric symbols of medieval Europe and continued in use into the 17th century. They are still used today for primarily decorative purposes. In fact, Romans numerals were not invented by the Romans, but date back to the Etruscans, a civilization found on the Italian peninsula between the seventh and fourth centuries BC, disappearing with the rise of the Roman Empire. The oldest of the Roman numerals, the symbols I, V, and X (1, 5, and 10), are probably prehistoric in origin and stem from the custom of cutting notches on tally sticks [1]. Some Roman numerals are given in Table 2.1, and an example of the multiplication process using Roman numerals is given in Table 2.2.

2.3 FURTHER DISCUSSIONS OF ZERO

The idea of numbers such as 1, 2, and 3 developed long before the concept of zero. This was largely because of a lack of need in early human societies. Consider, for instance, the case for counting fruit. While it is commonplace to say we have three apples, it is unlikely we would say we have zero apples. In such a case, we would say we do not have any apples. From

TABLE 2.1 Roman Numerals

Decimal	Roman	Decimal	Roman	Decimal	Roman
1	I	8	VIII	15	XV
2	II	9	IX	16	XVI
3	III	10	X	17	XVII
4	IV	11	XI	50	L
5	V	12	XII	100	C
6	VI	13	XIII	500	D
7	VII	14	XIV	1000	M

TABLE 2.2 A Multiplication Using Roman Numerals (CCLXV × XXXVIII)

Roman Notation	Decimal Notation
CCLXV	$100 + 100 + 50 + 10 + 5 = 265$
XXXVIII	$10 + 10 + 10 + 5 + 3 = 38$
CCLXV × XXXVIII	265×38
CCLXV × III = CCCCCC LLL XXX VVV	265×38
$\qquad$ = D CC L XXXX V $\qquad$ ... (1)	$= 265 \times 30 + 265 \times 8$
CCLXV × V = CCCCCCCCCC LLLLL XXXXX VVVVV	$= 7950 + 2120 = 10{,}070$
$\qquad$ = M CCC XX V $\qquad$ ... (2)	
CCLXV × XXX = MMMMM DDD CCC LLL ... (3)	
CCLXV × XXXVIII = (1) + (2) + (3)	
= MMMMMMM DDDD CCCCCCCC LLLL XXXXXX VV	
= MMMMMMMMMMLXX	
= $\bar{\text{V}}$MMMM D CCC CC L XX	
= $\bar{\text{V}}$MMMM DD L XX	
= $\bar{\text{V}}$MMMMM L XX	
= $\bar{\text{X}}$ L XX	

this vantage point, it is easy to see how societies got along for millennia without ever developing the concept zero. Even with the development of number systems, the concept of zero proved to be elusive.

The early Babylonians, despite their positional number system, did not invent the concept of zero, but rather represented an empty value by a blank space. Successive blank spaces on a clay tablet were ambiguous, and individual writing idiosyncrasies could also be misleading as to the presence or absence of a blank space between cuneiform wedges.

From the second century BC, the Babylonians finally decided to eliminate the blank space ambiguity by creating a symbol that served as a space holder for empty powers of the base. The symbol, never considered a numeral in itself, was two small slanted wedges.

The use of zero as a real value probably first appeared in India around the fifth century. *Lokavibhaga*, or "The Parts of the Universe," is the oldest known text to use a decimal place-value system that included a zero. This Jain cosmological text, dated 458 AD, uses Sanskrit words for numerals, including the word *shunya* for zero, meaning "void" or "empty" [5].

Aryabhata (476–550) was a renowned Indian mathematician and the author of several treatises on mathematics and astronomy. His best known work is *Aryabhatiya*, a text that summarizes Indian mathematical knowledge during the fifth and sixth centuries. Around 498 AD, Aryabhata used the letters of the traditional Sanskrit alphabet to record numbers. Some historians and mathematicians believe that his place-value system might be the origin of the modern decimal-based place-value notation, and that while he did not use a symbol zero, he certainly understood the concept [1, 6].

The appearance of a symbol for the digit zero, a small circle, was found on a stone inscription at the Chaturbhuja Temple at Gwalior in India, dated 876 AD. There are also many documents on copper plates, with a similar small circle in them, dated as far back as the sixth century AD, but some historians doubt their authenticity or true meaning.

While the Chinese had a positional number system that dated back to at least the second century BC, it was only much later, around the eighth century AD, when Buddhist missionaries brought the use of zero to China. Until that time, the Chinese left empty squares on their *shunya zi* to designate zero.

Before the rise of the Arab (Islamic) Empire, the decimal place-value number system, which originated in India, had already begun moving westward from the seventh century AD. As a result of the great wealth accumulated through their conquests, the Arabs, trading far and wide, became "cultural middlemen" throughout the Middle East, setting up Islamic centers in great cities such as Baghdad. They were eager to absorb the learning of other great cultures such as the Chinese, Indians, and Persians, as well as the Greeks and Romans.

In the year 773 AD, an Indian brought writings on astronomy to the court of the caliph of Baghdad. The book was translated from Sanskrit to Arabic. One of the Persian scholars to read this book was the famous Al-Khwarizmi, perhaps the greatest mathematician of his time. Based on his studies of Indian mathematics, he wrote a book entitled *Algorithmus* around the year 830, explaining the use of Indian numerals. In this way, the Indian place-value number system and the concept of zero was introduced to the Arabic world.

As Islamic and Christian cultures collided in the West, what had become known as Arab numbers gradually supplanted the use of Roman numerals and the abacus around the time the Renaissance era had begun in Europe. The place-value decimal number system we use today can therefore be traced directly to India. The positional number system and the concept of zero were the groundbreaking work of ancient Indian mathematicians [1, 2].

There is another story about the invention of zero. The oldest zero in the New World was found in Mayan civilization, dating back at least to the first century BC. The numeral is part of their *vigesimal* positional notation and was an independent invention. One prominent example of the Mayan zero is the Mesoamerican or Mayan Long Count Calendar developed by several pre-Colombian civilizations. The oldest known example of the calendar dates from 36 BC and comes from Mexico. The calendar used zero as a placeholder within its vigesimal (base-20) and base-18 place-value number systems. The Long Count Calendar identifies a date by counting the number of days since a mythical creation date corresponding to August 11, 3114 BC, in the Gregorian calendar [3]. In Mesoamerican numerals, a dot represented a 1, a bar represented a 5, and a snail shell-like symbol was used to represent the zero [4].

The numerals of the Long Count Calendar were no longer in use after the Spanish conquered the Yucatan Peninsula. Although early in its appearance, the Mayan zero did not spread to any location in the Old World, and consequently, the Mayan number system and zero did not influence the development of number systems elsewhere [3].

Over the millennia, cultures all around the world have experimented with number representation in many various ways trying to represent numbers. Many converge in their adoption of the decimal system, no doubt heavily influenced by the use of our 10 finger-digits to count in an earlier age. In those societies where positional notation was invented, one condition they share in common was the use of an abacus-like counting device, which seems to have predisposed them to adopting place-value in their written notation. It was in India that all of the factors came together to allow their mathematicians to invent a number system that was not only positional, but also had zero as a number. It was a long journey to get there, and the road was not straightforward, but this number system opened the door to the world of modern mathematics.

REFERENCES

1. Georges Ifrah, *The Universal History of Computing*, John Wiley & Sons, New York, 2001.
2. D. E. Smith, *History of Mathematics* (Vol. II), Dover Publications, New York, 1958.
3. Wikipedia, Mesoamerican Long Count Calendar, http://en.wikipedia.org/wiki/Mesoamerican_long_count_calender.
4. Wikipedia, Mesoamerica, http://en.wikipedia.org/wiki/Mesoamerican.
5. Wikipedia, Lokavibhaga, http://en.wikipedia.org/wiki/Lokavibhaga.
6. Wikipedia, Aryabhata, http://en.wikipedia.org/wiki/Aryabhata.
7. D. Smeltzer, *Man and Number*, Dover Publications, New York, 2003.
8. K. Menninger, *Number Words and Number Symbols*, Dover Publications, New York, 1992.
9. Wikipedia, Brahmi Numerals, http://en.wikipedia.org/wiki/Brahmi_numerals.

CHAPTER 3

Rational and Irrational Numbers

3.1 APPEARANCE OF FRACTIONS

Counting is the simplest and most fundamental operation using numbers. However, just counting was insufficient and inconvenient for people even in ancient civilizations. People at some early stage of their civilizations realized that arithmetic operations such as addition, subtraction, multiplication, and division were necessary and very important. It is probably more than 4000 years ago that people in Egypt, Mesopotamia, and other ancient civilized regions could already manipulate arithmetic operations using their number systems, although it is difficult to specify when and where people first discovered such techniques for calculations.

Among the four arithmetic operations, multiplication and division were difficult compared with addition and subtraction. People invented doubling and halving operations a long time ago. These and similar operations were convenient for their lives, and made multiplication and division easier. Consequently, the fractions 1/2, 1/4, 1/8, 1/16, and so on became commonly used numbers in addition to natural numbers. Furthermore, the use of reciprocals of integers, such as 1/2, 1/3, 1/4, 1/5, 1/6, and so on, also became common and of great importance. The reciprocal of each nonzero natural number is called a *unit fraction*. They noticed that multiplications of unit fractions by integers were also useful for calculations. In this way, fractions appeared in ancient civilizations [7].

Ancient Egyptians used special symbols representing fundamental unit fractions and some other fractions (e.g., 2/3). These symbols appeared in several mathematical tablets and papyri found in Egypt [10]. Akhmim wooden tablets and Cairo wooden tablets are two ancient Egyptian documents that contain, for example, descriptions about multiplications by fractions 1/3, 1/7, 1/10, 1/11, and 1/13. The following calculation of fractions also appeared in the tablets:

$$1/2 + 1/4 + 1/8 + 1/16 + 1/64 + 5/320 = 1$$

Historians suggest that these tablets were probably inscribed at the beginning of the Egyptian Middle Kingdom around 1950 BC [3]. These are now held in the Egyptian Museum in Cairo.

In the British Museum in London, the Rhind Papyrus (also called Rhind Mathematical Papyrus) is displayed. It is approximately 5 m long and 33 cm wide, one of the oldest existing texts of Egyptian mathematics. A Scottish lawyer, Alexander H. Rhind (1833–1863), purchased it in 1858 in Egypt [8]. It was named the Rhind Papyrus after him. It was copied by a scribe named Ahmes around 1650 BC. In the first paragraph of the papyrus, Ahmes presents that it is copied from an ancient copy made during the 12th dynasty of Upper and Lower Egypt (c. 1985–1795 BC) [3, 8].

The first part of the Rhind Papyrus contains a list of the fractions $2/n$ for odd n from 3 to 101. The following are examples in the list:

$$2/3 = 1/2 + 1/6$$

$$2/5 = 1/3 + 1/28$$

$$2/7 = 1/4 + 1/28$$

$$2/9 = 1/6 + 1/18$$

$$2/15 = 1/10 + 1/30$$

$$2/101 = 1/101 + 1/202 + 1/303 + 1/606$$

The second and third parts of the Rhind Papyrus consist of geometry problems, and 84 problems with the solutions, respectively. Below is an example in the third part:

Problem 3.1

Let the sum of 2/3 and 1/10 of an unknown quantity be 10. Calculate the unknown quantity.

SOLUTION

Using modern notation, $(2/3 + 1/10)x = 10$, where x is the unknown quantity. Then $x = 300/23$.

The Moscow Papyrus (also called Moscow Mathematical Papyrus) is now held in the Puskin State Museum of Fine Arts in Moscow [9]. It is approximately 5.5 m long and $3 \sim 7.7$ cm wide, one of the oldest Egyptian mathematical texts in existence. The Moscow Papyrus probably dates to the 11th dynasty of Upper and Lower Egypt (c. 1850 BC). It contains 25 problems of arithmetic, algebra, and geometry. The following are examples of the problems:

Problem 3.2

Let the sum of 1 and a half of an unknown quantity and 4 be 10. Calculate the unknown quantity.

SOLUTION

Using modern notation, $(1 + 1/2)x + 4 = 10$, where x is the unknown quantity. Then $x = 4$.

Problem 3.3

Let $1/2 + 1/4$ of the square of an unknown quantity be 12. Calculate the unknown quantity.

SOLUTION

Using modern notation, $(3/4)(3/4)x^2 = 12(3/4)$, where x is the unknown quantity. Then $x = 4$.

3.2 RATIONAL NUMBERS

The set of *rational* numbers is usually denoted by $\mathbf{Q}$, and the elements of this set have the property that they can be represented as a *ratio* of two integers (hence the name *rational*). These numbers are, in a sense, *measurable*, and they include whole integers, fractions, and their together negative counterparts. The set of rational numbers is equivalent to the set of multiplications of unit fractions and integers. That is, we can write

$$Q = \{p/q \mid p \text{ is an integer and } 1/q \text{ is a unit fraction}\}$$

A rational number is one that can be *named* number, in that we can say *seven, three-fourths, 0.3*. With this, any rational number can be placed in its exact (measured) position on the real number line (see also Chapter 13). Rational numbers can be represented using decimal notation where the representation is either finite (e.g., two-fifths is 0.40) or infinite, but repeating (e.g., three and two-sevenths is 3.285714285714285714...). Historically the decimal fraction notation (e.g., 3.285714285714285714...) appeared much later than fraction notation (e.g., 3 + 2/7 or 23/7). The use of decimal fraction notation became common in the 16th century (see more in Chapter 10). Since a repeating decimal can be written algebraically as a fraction, it follows that any repeating decimal must be a rational number. Specifically, to prove that any repeating decimal represents a fraction of two integers, let us examine a number x whose decimal representation is of the form $i_1 i_2 \cdots i_k . d_1 d_2 \cdots d_m abcdabcdabcd \cdots$, where the digits i and d represent the integer part, and the nonrepeating portion of x and $abcd$ is the repeating pattern. Algebraically, $x = r + y$, where $r = i_1 i_2 \cdots i_k d_1 d_2 \cdots d_m / 10^m$, which already is a ratio of two integers, is a rational number. So, we just have to show that the number $y = 0.00...0abcdabcdabcd...$ is also a ratio of two integers. We multiply assign to g the repeating pattern value of $abcd$, and pick the smallest h such that $10^h > abcd$; i.e., h is the pattern length. Now, $y = abcd/(10^h - 1)$ is a ratio of two integers; hence, it must be rational. Since the set of rational numbers is closed under addition (i.e., a sum of two fractions is a fraction), the original number x is rational.

Let us examine the above formula on the repeating decimal 0.999.... The pattern length here is just one digit. Therefore, we have $9.0/(10^1 - 1) = 1.0$. This is unexpected, because it implies that there are two different representations for the same number, that is, 1. The first is 1.0 and the second is 0.999.... Intuitively, we know that 0.999... is three times 0.333... = 1/3. So the fact that 0.999... = 1.0 is, at least, not counterintuitive. However, it is concerning that the same (rational) number can have two representations. We can prove that 0.999... = 1.0 in another way. Let us solve the following system of equation(s), 0.999... = x, with one equation and one unknown.

Algebraically, we can multiply the two sides of an equation by a number and maintain equality. In this case, let the multiplier be 10. So, we have $9.999\ldots = 10x$. Now, if we subtract the first equation from the second, we have $9.0 = 9x$, so we get our final answer: $x = 1.0$.

3.3 IRRATIONAL NUMBERS

Pythagoras (c. 570–540 BC) is one of the most famous Greek philosophers and mathematicians. The *Pythagorean theorem* is a relation among the side lengths of a right triangle. Let a, b, and c be the side lengths of a right triangle, where c is the length of the side opposed to the right angle. The Pythagorean theorem can be expressed as the following equation:

$$a^2 + b^2 = c^2.$$

The Pythagorean theorem was probably known much before Pythagoras [14], but it is said that Pythagoras first proved the theorem. There exist many different proofs of the Pythagorean theorem. It can be proved algebraically or geometrically, e.g., by using the drawings in Figure 3.1.

From the Pythagorean theorem, the Greek mathematicians, probably including Pythagoras, realized that there exist numbers that cannot be measured by a rational number. The most obvious example was the length of the diagonal of a one-by-one square, which is $\sqrt{2}$ from the Pythagorean theorem. More generally, if n is a positive integer and not a square number, then $\sqrt{n}$ is not a rational number. The proof of this fact is given in *Theaetetus* (c. 360 BC) written by the Greek philosopher and mathematician Plato (427–347 BC). They noticed that these numbers cannot be represented by a ratio of two integers. Arabic mathematicians treated these numbers as algebraic objects. Hindu and other mathematicians were also aware of the existence of such numbers and that they are much more mysterious than the rational numbers, but none of the early mathematicians grasped the full meaning and magnitude of this set.

Pythagoras wanted to believe that all numbers were rational (could be written as a fraction or be measured). Hippasus (fifth century BC), who was Pythagoras's student, actually showed that the square root of 2 cannot be a ratio of two integers. It is believed that he used a geometrical (and not an algebraic) argument. Pythagoras was caught in a dilemma: what to do with a number that was not rational, i.e., *irrational*. Despite Hippasus's proof, he would not accept the existence of irrational numbers. Since Pythagoras could not disprove the existence of irrational numbers, he had

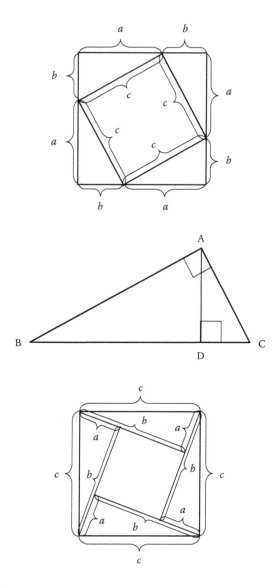

FIGURE 3.1   Three different proofs of the Pythagorean theorem.

Hippasus thrown overboard and drowned! An unfortunate footnote to this is that $\sqrt{2}$ is often called *Pythagoras's constant*.

Let us present a simple (non-constructive [6]) proof that there can exist no rational number x whose square is 2. Suppose such a number existed, i.e., $x = m/n$, where n is not 0, m/n is simplified to the lowest terms (i.e., m/n is an irreducible fraction), and $x^2 = 2$. It follows that $(mm)/(nn) = 2$. Therefore, both m and n cannot be even numbers—at least one of them must be odd *since m/n is irreducible.* From the equation, we can derive $n^2 = 2m^2$, which is an even number. This implies that m must be an even number, e.g., $m = 2k$, where k is one-half of m. Now, let us substitute for m. We have $(mm)/(nn) = 2 = (2k)^2/(nn) = 4k^2/(nn)$. So, $2nn = 4k^2$, or equivalently, $nn = 2k^2$. Therefore, nn must be even, which makes n even. But both m and n could not be even (since m/n is irreducible), and so we have a contradiction. This means that our assumption that $\sqrt{2}$ is a ratio of two integers (rational number) is not correct.

While the majority of tasks of daily life involve rational numbers, the irrational numbers are just as important, and as will be shown later, there are much more irrational numbers than we can count (see Chapters 13 and 14).

REFERENCES

1. Wikipedia, Real Number, http://en.wikipedia.org/wiki/Real.
2. J. C. Tweddle, Wierstrass's Construction of the Irrational Number, *Math Semester*, 58, 47–58, 2011.
3. D. E. Smith, *History of Mathematics* (Vol. II), Dover Publications, New York, 1958.
4. Mathworld, http://mathworld.wolfram.com/IrrationalNumber.html.
5. Wikipedia, Irrational Number, http://en.wikipedia.org/wiki/Irrational_number#cite_note-28.
6. A. George and D. Velleman, *Philosophies of Mathematics*, Blackwell, 2002.
7. Wikipedia, Fraction, http://en.wikipedia.org/wiki/Fraction.
8. Wikipedia, Rhind Papyrus, http://en.wikipedia.org/wiki/Rhind_mathematical_Papyrus.
9. Wikipedia, Moscow Papyrus, http://en.wikipedia.org/wiki/Moscow_Papyrus.
10. Wikipedia, Akhmim Wooden Tablets, http://en.wikipedia.org/wiki/Akhmim_wooden_tablets.
11. G. Ifrah, *The Universal History of Computing*, John Wiley & Sons, New York, 1981.
12. R. G. Robins and C. C. Shute, *The Rhind Mathematical Papyrus: Ancient Egyptian Text*, British Museum Publications, London, 1987.
13. Mathematics in Egyptian Papyri, http://www-history-mcs.st-andrews.ac.uk/HistTopics/Egyptian_papyri.html.
14. B. L. van der Waerden, *Geometry and Algebra in Ancient Civilization*, Springer-Verlag, Berlin, 1983.

CHAPTER 4

Prime Numbers

Prime numbers, defined as natural numbers (also called counting numbers), divisible by themselves and 1 only, have long enjoyed a special place of mystery for both advanced mathematicians and school children alike. The smallest prime number is 2, and it is the only even prime number. For it seems that the moment, as a civilization, we could perform the arithmetic operation of division, the primes have presented themselves as special, and to this day, many questions of primes are yet to be answered.

For instance, can we consider the primes "building blocks" of the natural numbers (as Jon Keating, professor of mathematical physics at the University of Bristol, asked)? Can we say that from primes all other numbers can be constructed? How many, and in what way? Consider that a typical layperson, given the task of identifying a large prime, will likely do so by first producing a guess and dividing it by all the prior prime numbers less than or equal to its square root. Depending on the size of the provided prime (and the memory of the individual), this operation can be quite daunting. For centuries now, mathematicians, called *number theoreticians*, have sought to answer exactly these questions.

In the following sections, we will examine what progress we have made over time with understanding primes. We will also see that there are questions for which we have answers, and there are many which remain open and are under study even today.

4.1 THE STORY OF PRIME

The story of prime numbers likely begins in unrecorded history, when the idea of groups being broken into smaller groups first occurred to mankind. We can see written accounting for prime numbers in an ancient mathematical papyrus that appeared in Egypt around 1650 BC. Ancient Egyptians demonstrated perhaps some awareness of primality. It is about that time that a scribe named Ahmes recorded a possible understanding of the significance of primes within what is now known as the Rhind Mathematical Papyrus [7] (see also Chapter 3). Within the papyrus, an exercise of sorts is performed listing a table of fractions as composites of factors, and it is within this table the notable separation of prime-based fractions is done.

While the exact implication of the separation and the extent of the Egyptians' knowledge of primes at the time are difficult to ascertain from this evidence, the archeological finding is fascinating. It is sometime later, about 500 to 300 BC, that the first formal descriptions of primes and perfect numbers surface within a group known as the Pythagoreans, followers of Pythagoras, who is famous for the Pythagorean theorem: $a^2 + b^2 = c^2$.

Perfect numbers are numbers that are the sum of their proper factors. For example, 6 is a perfect number, as its factors 3, 2, and 1 add to 6. Euclid provided a more proper definition in a subsequent paragraph ([6] Book 7, Definition 22).

An interesting note on the Pythagoreans: while we are perhaps familiar with their early interest in mathematics, it is not entirely secular or academic in nature. In fact, the Pythagoreans seemed to be of the opinion that numbers were entities unto themselves, and worthy of consciousness and even divinity. Although the Pythagoreans may be our earliest documented description of primes, it was not until Euclid's *Elements* at the end of that period that we are given a more comprehensive understanding (complete with definitions, propositions, and proofs).

Euclid's *Elements* is widely regarded as the single most important ancient contribution to the whole of mathematics, engineering, and science. Its codification and format are outstanding, where definitions are used axiomatically, and propositions grow from those definitions and earlier propositions. It consists of 13 books (volumes) that are a collection of the known mathematics at that time. It had been the standard textbook for mathematics education for some time (see also Chapter 5).

It is also worth noting that it is of contentious authorship: it is attributed to Euclid, but it is presumed to contain the contemporary sum of mathematics for his time. While how much of the content can be attributed to Euclid himself is a moot point, the existence of such well-organized documentation is of interest to us.

We can begin by giving the *Elements'* definition of prime: "A prime number is that which is measured by a unit alone" ([6] Book 7, Definition 11). Euclid's definition is characteristically Greek, harkening from the Pythagoreans in the reference of the unit. Euclid states earlier in the *Elements* that a number is comprised of units, and the unit is simply that which by its existence is "called one" ([6] Book 7, Definitions 1 and 2). Simply put, a prime number has only 1 as its natural divisor, without a remainder. With this definition, Euclid makes three important propositions:

1. If a number is a product of two numbers and has a factor of a prime, then one of the two numbers has a factor of the prime.

2. Any nonprime number has a prime number as its factor.

3. Any number is either prime or divisible by a prime.

Summarized, these statements become: "any integer can be expressed as a unique ordered product of primes," which is typically known and understood to be the *fundamental theorem of arithmetic*. Although Euclid proposes a proof, it is worth deferring until later. Euclid makes one further celebrated contribution to primes with a proof that there are infinitely many prime numbers.

Proof by Contradiction

Consider a set of primes, $a_0 a_1, \cdots, a_n$, and $m = \prod_{i=0}^{n} a_i$.

For a number $m + 1$ two cases are considered against the propositions that lead to the fundamental theorem of arithmetic:

If $m + 1$ is prime, then there are at least $n + 2$ primes (since it would be $m + 1$ in the set).

If $m + 1$ is not prime, then there are at least $n + 1$ primes, as some prime must be a factor within it, as $a_0 a_1, \cdots, a_n$ are factors of m (and not $m + 1$, as division by any of the set $a_0 a_1, \cdots, a_n$ would leave a remainder of 1).

Another definition of scholarly interest is that of so-called *perfect numbers* (whose significance will soon be revealed). In the ancient times, only the first few perfect numbers were known. As of 2013, there are only 48 known perfect numbers, and it is still unknown whether the set of even perfect numbers is finite or infinite, or if there exist any odd perfect numbers. The following numbers are examples of perfect numbers:

$$6 = 3 + 2 + 1$$

$$28 = 14 + 7 + 4 + 2 + 1$$

$$496 = 248 + 124 + 62 + 31 + 16 + 8 + 4 + 2 + 1$$

As should be expected, *Elements* provides a relationship between primes and perfect numbers, specifically that $2^n (2^{n+1} - 1)$ is a perfect number if $2^{n+1} - 1$ is a prime. Considering the limits of the computational techniques available at the time, this relationship is remarkable and will figure predominantly in mathematical works to come.

The Pythagoreans again enter history when Nicomachus (c. 60–c. 120 AD) reported a technique from the third century BC by a scholar named Eratosthenes (276–194 BC). The technique is called (appropriately enough) the sieve of Eratosthenes. Eratosthenes, a contemporary of Archimedes, developed an iterative and mechanical technique for sieving the natural numbers for primes (the term *mechanical* is used purposely, in consideration of a famous letter from Archimedes to Eratosthenes arguing that certain mathematical problems are better handled mechanically, i.e., without thinking too much).

Eratosthenes is perhaps best known for calculating the circumference of the earth, its tilt (relative to the sun), and its distance from the moon. He was also a noteworthy playwright and a poet, and furthered (if not founded the modern form of) the subject of geography. Unfortunately, we only know of his life and works through other sources, the originals lost to history.

From Nicomachus's description, however, it is possible to accurately reconstruct his simple solution to the problem of finding prime numbers.

Table 4.1 demonstrates the algorithm for primes between 1 and 100. The procedure is to sieve the numbers less than or equal to 100 by first eliminating multiples of 2 (▱), then 3 (◫), 5 (⊠), and 7 (⊟). The

TABLE 4.1 Finding Primes by the Sieve of Eratosthenes

	2	3	4	5	6	7	8	9	10
11	12	13	14	15	16	17	18	19	20
21	22	23	24	25	26	27	28	29	30
31	32	33	34	35	36	37	38	39	40
41	42	43	44	45	46	47	48	49	50
51	52	53	54	55	56	57	58	59	60
61	62	63	64	65	66	67	68	69	70
71	72	73	74	75	76	77	78	79	80
81	82	83	84	85	86	87	88	89	90
91	92	93	94	95	96	97	98	99	100

The *sieve of Eratosthenes* has two basic steps:

1. List all the natural numbers from 1 to n within which the primes of interest must lie.
2. Then, starting at 2, eliminate all multiples within the list. Repeat this step for all numbers from 2 to $\sqrt{n}$.

When the process is complete, the numbers that remain in the list are prime.

maximum necessary step to consider is = 10, and the numbers 4, 6, 8, 9, and 10 were eliminated by earlier factors (2 and 3).

It was about 1650 that our next pioneer of prime numbers, Pierre de Fermat (1601–1665), stated that for a prime p and an integer a such that p and a are relatively prime, $a^{p-1} \equiv 1 \pmod{p}$ or $a^{p-1} - 1$ is divisible by p. It was left to Euler to formulate the proof, but this, his "little" theorem, would become the basis for public-key cryptography—an essential tool of modern life. Fermat has been called the greatest amateur mathematician, although a number of his theorems were stated with no proof [1].

Relatively prime: Two numbers are considered relatively prime (also called coprime) if their greatest common divisor is 1.

In 1736 Leonhard Paul Euler (1707–1783) was the first to document a proof of Fermat's little theorem. Here we give a proof of Fermat's little theorem using modern algebra [2].

Proof

First, we define Euler's totient function (also called the Euler phi-function). For n within positive natural numbers, $\varphi(n)$ is the number of integers k coprime to n such that $1 \leq k \leq n$. This is equivalent to $p - 1$ if n is a prime p. Let $Z_n^* = \{k \mid k$ is a positive integer coprime to n and less than $n\}$, and let the multiplication of any pair of elements of Z_n^* be the multiplication of the elements *modulo n*. Then Z_n^* is a multiplicative group of order $\varphi(n)$, and for any a in Z_n^*, $\varphi(n)$ is divisible by the order of a (from Lagrange's theorem), where the order of a is the smallest positive integer m such that $a^m = 1$ in Z_n^* [2, 5]. If n is prime (say $n = p$), then $\varphi(n)$ is $p - 1$. Hence, $a^{p-1} \equiv 1 \pmod{p}$. ∎

As alluded to within the explanation of Fermat's little theorem, Euler's proof (essentially the same as the proof above) gives the mathematical blueprint for the public-key encryption. If n is the product of two primes p_1 and p_2, then $\varphi(n) = (p_1 - 1)(p_2 - 1)$ and you have an interesting function dependent on the prime factors of n. Given two large primes p_1 and p_2, it is easy to multiply p_1 and p_2, but given only the number n, it is extremely difficult to factor for the primes (see also Chapter 30).

Euler went on to contribute two other conjectures regarding prime numbers. The first result, later proven by Pafnuty Chebyshev (1821–1894), is that for any integer greater than or equal to 2, there exists a prime between itself and twice itself. The second is drawn from an equation from the Pythagoreans to generate primes, Euler's own $x^2 - x + 41$. The discovery of other polynomials capable of generating primes continued with the work by Legendre and many others.

A few years later we turned our attention to the idea that numbers can be represented by sums of primes, similar to the building block idea in the introduction to this chapter. In a letter from Christian Goldbach (1690–1764) to Euler in 1742 is written what is now known as *Goldbach's conjecture*, considered one of the great open problems remaining in prime number theory. Simply stated in its modern form, it is that every even integer greater than 2 can be expressed as the sum of two primes.

4.2 THE PRIME NUMBER THEOREM

It is at this point in history where things get a little more contentious, as two great mathematicians work toward a common problem, that of $\pi(n)$, or the *prime number theorem*.

In 1796 Adrien-Marie Legendre (1752–1833) published a book, *Essay on the Theory of Numbers*, within which he conjectured the distribution of primes could be described as

$$\pi(n) \sim \frac{n}{\ln n - B}$$

for some value B (Figure 4.1).

This remarkable result, however, would be overshadowed by a young Gauss (Johann Carl Friedrich Gauss, 1777–1855), and it would not be the first time Gauss claimed a better or earlier discovery.

Two stories, perhaps more legend than fact, attempt to hint at the genius that was to come in Gauss. It is said that at the age of 3, Gauss noted an error in wages paid to workers by his father, and at the age of 7, tasked to sum the numbers 1 to 100 by his instructor, he added them by summing pairs ({1, 100}, {2, 99}, …, {50, 51}), although this story is also attributed to the great Albert Einstein (1879–1955).

There is no doubt, however, that as a teenager, Gauss had demonstrated extraordinary abilities. Following Euclid's techniques of geometry (using

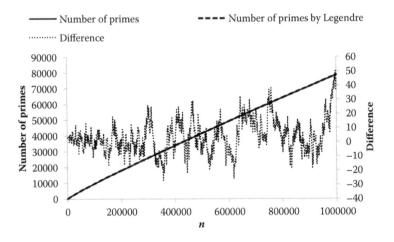

FIGURE 4.1 The numbers of primes less than 1,000,000 and the conjectured number of primes by Legendre ($B = 1.08366$).

only a straightedge and a compass for construction), he constructed a 17-sided heptadecagon, well past the known constructions for polygons with sides of multiples of 2, 3, and 5. He went on to add many other similar geometries to his repertoire, all of the form $2^{2^n}+1$ (which we know as *Fermat primes*). From about ages 15 to 18, during his study at the Brunswick Collegium Carolinum, Gauss began to think about the distribution of primes. Similarly to Legendre, Gauss began his study observationally—he is said to have spent an idle 15 minutes counting the primes in blocks of a thousand, and when his survey neared a million, he proposed what would later be known as the prime number theorem, a bounding function for the distribution of primes:

$$\pi(n) \sim \frac{n}{\ln n}$$

which was then later revised to

$$\pi(n) = \text{Li}(n), \text{ where } \text{Li}(n) = \int_2^n \frac{dn}{\ln n}$$

Two numbers, x and y, are said to be congruent modulo z if their difference, specifically $x - y$, is divisible by another number z. The notation for the above is

$$x \equiv y \ (mod \ z)$$

Gauss would go on to make many other contributions to mathematics and science, but another of his works, this one at age 24, is of great interest to the study of primes. In fact, it would unify number theory at his time. The principle Gauss introduced is called congruence.

The problem he set out to solve was stated, although not proven, by Legendre, and is known as the *quadratic reciprocity theorem*. Briefly, the theorem states if p and q are distinct odd primes, not both congruent to 3 modulo 4 (both don't have a remainder 3 when divided by 4), then either both $x^2 \equiv p \ (\text{mod } q)$ and $x^2 \equiv q \ (\text{mod} p)$ are solvable, i.e., have a solution, or neither is solvable [2].

In 1831 August Ferdinand Möbius (1790–1868), famous for the later invention of the *Möbius strip*, invented a function, $\mu(n)$, called the Möbius function, formulated as follows: it will produce the value 0 for any n that is a multiple of squares such as 4, 9, 16, and 25; the value 1 for an n with an even number of distinct primes as its factors or $n = 1$; and −1 for integers that have an odd number of distinct primes as its factors.

While its usefulness is not readily apparent, it demonstrates the breadth of the participants involved in solving problems related to primes. This function, paired with another one called *Marten's function*, proves useful as an alternate path of proving the *Riemann hypothesis*.

In 1859 Georg Friedrich Bernhard Riemann (1826–1866) introduced what is today worth a million dollars (if proved correctly), the so-called Riemann hypothesis. It is considered one of the greatest problems mathematics has yet to solve [3, 4].

First, it is worth knowing a bit about Riemann. Like Gauss, Riemann demonstrated a keen ability in his teenage years. One of his notable feats in his youth was responding to the challenge of one of his instructors by reading Legendre's 859-page book, *Theory of Numbers*; not only did he read it in a week, but he was able to answer questions regarding its contents 2 years later.

While in Riemann's short life he offered a great deal to mathematics (much of it only known from his private papers after his death), it is the Riemann hypothesis, a brief remark made in a paper regarding an attempt to solve the question of the distribution of primes from another direction, specifically from an infinite series. The hypothesis progresses as such:

$$\left(\frac{1}{1^s}\right)+\left(\frac{1}{2^s}\right)+\left(\frac{1}{3^s}\right)+\cdots+\left(\frac{1}{n^s}\right)+\cdots$$

First, consider an infinite series, familiar from Euler, called by Riemann the *zeta function*:

$$\zeta(s)=\sum_{n=1}^{\infty}\frac{1}{n^s}$$

From here, two techniques can be used to understand its progression to our more common zeta function form for the Riemann hypothesis. Should the reader be familiar with infinite geometric series, it requires only substituting infinite products producing a series of the form (it converges to $1/(1 - p^{-s})$):

$$1 + \frac{1}{p^s} + \frac{1}{p^{2s}} + \cdots + \frac{1}{p^{ks}} + \cdots$$

For the second technique, consider algebraic manipulation by dividing both sides by the terms within the series, performing an operation analogous to the sieve of Eratosthenes. The result of both will be a series of the form:

$$\zeta(s) = \frac{1}{1 - \frac{1}{2^s}} \times \frac{1}{1 - \frac{1}{3^s}} \times \frac{1}{1 - \frac{1}{5^s}} \times \frac{1}{1 - \frac{1}{7^s}} \times \frac{1}{1 - \frac{1}{11^s}} \times \cdots \times \frac{1}{1 - \frac{1}{3^s}} \times \cdots$$

We then produce the zeta function of interest:

$$\zeta(s) = \Pi_p (1 - p^{-s})^{-1}$$

The hypothesis, then, is this: the roots of the zeta function would determine the magnitude of the difference between Gauss's bounding function Li(x) and the true $\pi(x)$. Proof of this hypothesis is therefore both figuratively and literally the million dollar question within mathematics.

There have been interesting developments within the study of primes in the last century as well. For example, in 1919 Viggo Brun (1885–1978) proved that when the reciprocals of successive twin primes are added, their sum converges to a specific value called *Brun's constant*. While it is known that the sum of all the reciprocals of primes diverges, it is interesting that for twin primes it converges. It is known that twin primes, with the exception of the first ({3, 5}), are of the form $(6n - 1, 6n + 1)$.

In 1960, a Polish mathematician, Waclaw Franciszek Sierpiński (1882–1969), proved there are infinitely many nonprime odd integers k such that $k \times 2^n + 1$ is composite. The smallest known *Sierpiński number* is 78,557, for which all numbers of the form are divisible by 3, 5, 7, 13, 19, 37, or 73, and work is currently under way with a distributed computing project to attempt to prove it the smallest [8].

Three years later, another Polish-born mathematician, Stanislaw Marcin Ulam (1909–1984), famous primarily for his work on the *Manhattan Project*, the *Monte Carlo method*, and many other contributions, was bored at the reading of a paper and created a doodle uncovering what is perhaps a remarkable pattern to certain primes—specifically, that when natural numbers are written in a spiral from the number 1, prime numbers tend to appear on diagonals with one another.

There continues to be research performed on primes, twin primes, and perfect numbers. Ever faster and more powerful computers continue to hint at mysteries locked within primes. From Dorin Andrica's 1985 conjecture that the gap between two prime numbers is $\sqrt{p_{n+1}} - \sqrt{p_n} < 1$ to Andrew Granville's (1962–) 2008 paper on prime number distribution patterns, the study of primes goes on [9].

REFERENCES

1. J. von zur Gathen and J. Gerhard, *Modern Computer Algebra* (2nd ed.), Cambridge University Press, Cambridge, UK, 2003.
2. D. R. Stinson, *Cryptography: Theory and Practice*, CRC Press, New York, 1995.
3. K. Sabbagh, *The Riemann Hypothesis: The Greatest Unsolved Problem in Mathematics*, Farrar, Straus and Giroux, New York, 2002.
4. J. Derbyshire, *Prime Obsession: Bernhard Riemann and the Greatest Unsolved Problem in Mathematics*, Joseph Henry Press, Washington, DC, 2003.
5. G. Birkhoff and S. MacLane, *A Survey of Modern Algebra* (3rd ed.), Macmillan, New York, 1965.
6. Euclid (author), T. L. Heath (trans.), and D. Densmore (ed.), *Euclid's Elements*, Green Lion Press, Santa Fe, NM, 2002.
7. R. J. Gillings, The Recto of the Rhind Mathematical Papyrus: How Did the Ancient Egyptian Scribe Prepare It? *Archive for History of Exact Sciences*, 12(4), 291–298, 1974.
8. Seventeen or Bust, A Distributed Attack on the Siepinski Problem, http://www.seventeenorbust.com/.
9. Wikipedia, Andrica's Conjecture, http://en.wikipedia.org/wiki/Andrica's.

CHAPTER 5

Euclid's Elements

Euclid (also called Eukleides or Euclid of Alexandria, c. 330–c. 260 BC) is one of the most prominent Greek mathematicians. He is best known as the author of *Elements*, which is the most influential treatise in the history of mathematics. However, there are no known records of the exact date and place of Euclid's birth, and little is known about his personal life. During the reign of the Pharaoh Ptolemy I Soter (323–283 BC) Euclid taught mathematics at Alexandria Library (the Mouseion) in Alexandria, Egypt. Euclid's *Elements* is considered the most comprehensive compilation of geometry, arithmetic, and number theory based on the ancient Greek works of Thales (c. 624–c. 546 BC), Pythagoras (c. 582–c. 497 BC), Plato (c. 427–c. 347 BC), Theaetetus (c. 417–369 BC), Eudoxus (c. 408–c. 347 BC), Aristotle (384–322 BC), Manaechmus (380–320 BC), and others [2, 4, 6, 9].

Euclid is often referred to as the *father of geometry*. It is thought that he received mathematical training in Plato's Academy in Athens. Then, he came to Alexandria in Egypt, which had been the Hellenistic center for some centuries. Many scholars worked and taught at the great library in Alexandria, and Euclid wrote *Elements* there, which was the most widely used textbook of all time until the 20th century. This treatise influenced the development of Western mathematics for more than 2000 years. Proclus (412–485 AD), another Greek philosopher, wrote an influential commentary on Euclid's *Elements*. According to this commentary, to the Pharaoh Ptolemy I Soter's request for an easy way of learning mathematics, Euclid replied, "There is no royal way to geometry" [2, 4, 5, 6].

Elements consists of 13 books. It is a collection of definitions, postulates, common notions, propositions, and mathematical proofs of the propositions. Books 1 through 4 are on plane geometry. Book 1 begins with 23 definitions, 5 postulates, and 5 common notions, and contains 48 propositions. We may consider that both postulates and common notions are axioms. The following are the *postulates* given in Book 1 [1]:

1. We can draw a straight line from any point to any point.

2. We can produce a finite straight line continuously in a straight line.

3. We can draw a circle with any center and radius.

4. All right angles are equal to one another.

5. One and only one line can be drawn through a point parallel to a given line.

(In the 19th century *non-Euclidean* geometry was introduced, in which the fifth postulate was removed.)

The following are common *notions* that are also given in Book 1. These are not specific geometrical properties, but general assumptions used in mathematics:

1. Things equal to the same thing are also equal to one another.

2. If equals are added to equals, the whole are equal.

3. If equals are subtracted from equals, the whole are equal.

4. Things that coincide with one another are equal to one another.

5. The whole is greater than the part.

All propositions in Books 1 through 4 are proved by graphical constructions using axioms or propositions proved earlier in the books. For example, Proposition 1 of Book 1 shows a graphical construction of an equilateral triangle on a given finite straight line. Proposition 47 of Book 1 is the *Pythagorean theorem*. Specifically, in Proposition 47, it is proved by a graphical construction that in any right-angled triangle, the area of the square whose side is the hypotenuse is equal to the sum of the squares whose sides are the two other sides. Since most of the propositions in Book 2 can be considered geometric interpretations of algebraic identities, Book 2 is called the book of *geometric algebra*. For example,

Proposition 5 of Book 2 is a geometric interpretation of the algebraic equation $(x + y)(x - y) = x^2 - y^2$.

Proposition 5 (of Book 2)

Let a straight line AB be cut into equal segments at C and into unequal segments at D, as shown below:

A____C__D_B

Then the sum of the space of the rectangle whose side lengths are AD and DB, and the space of the square whose side length is CD, is equal to the space of the square whose side length is AC. ▪

Book 3 begins with 11 definitions and contains 37 propositions. It studies the properties of circles. For example, Proposition 1 of Book 3 states how to find the center of a given circle, and Proposition 5 shows the property that if two circles cut one another, then they will not have the same center. Book 4 deals with problems about circles, including constructions of regular polygons with 4, 5, 6, and 15 sides. Book 5 begins with 18 definitions followed by 25 propositions about magnitudes, multiples, ratios, and proportions of numbers and line segment lengths. Book 6 contains 33 propositions. There are some applications of the results of Book 5 to plane geometry.

Book 7 deals with elementary number theory. It includes divisibility, prime numbers, and algorithms for finding the greatest common divisor (GCD). The *Euclidean algorithm*, which shows an efficient method for finding the GCD of two integers, is given in Propositions 1 and 2 of Book 7 and in Propositions 2 and 3 of Book 10. In Book 7, the algorithm is formulated for integers, whereas in Book 10, it is formulated for lengths of line segments. The Euclidean algorithm starts with a pair of numbers (positive integers) and forms a new pair of numbers that consists of the smaller number and the difference between the larger and smaller numbers. This process repeats until the numbers become equal. The resulting number is the GCD of the original pair of numbers. Euclid also gave a method (Proposition 34 of Book 7) for finding the least common multiple (LCM) of two integers.

In a modern textbook on elementary number theory or computer algorithms, the Euclidean algorithm is usually described in the following way.

Let r_0 and r_1 be a given pair of positive integers, where $r_0 > r_1$. The algorithm consists of performing the following sequence of divisions:

$$r_0 = q_1 r_1 + r_2, \quad 0 < r_2 < r_1$$

$$r_1 = q_2 r_2 + r_3, \quad 0 < r_3 < r_2$$

$$\vdots$$

$$r_{m-2} = q_{m-1} r_{m-1} + r_m, \quad 0 < r_m < r_{m-1}$$

$$r_{m-1} = q_m r_m$$

Then it is not hard to show that

$$\gcd(r_0, r_1) = \gcd(r_1, r_2) = \ldots = \gcd(r_{m-1}, r_m) = r_m$$

where $\gcd(a, b)$ means the GCD of a and b. Hence, it follows that $\gcd(r0, r_1) = r_m$.

The Euclidean algorithm was probably not discovered by Euclid. As D. E. Knuth states in his book [7], some scholars believe that the method was known up to 200 years earlier, and it was almost certainly known to Eudoxus [7]. B. L. van der Waerden suggested that Book 7 was derived from a textbook written by mathematicians in the School of Pythagoras [9]. Centuries later, the Euclidean algorithm was discovered independently in India and in China to solve Diophantine equations (see Chapter 6). Knuth calls the Euclidean algorithm the *granddaddy* of all algorithms, because it is the oldest nontrivial algorithm that has survived to the present day [7].

The original Euclidean algorithm was described only for natural numbers and geometric lengths, but in the 19th century it was generalized to other types of numbers such as modular arithmetic numbers and polynomials in one variable. Although the Euclidean algorithm is one of the oldest algorithms, it is still commonly used. The algorithm has many theoretical and practical applications. It is an important part of the construction of the RSA cryptography, a public-key cryptosystem most widely used in the security of electronic commerce on the Internet (see Chapter 30). It is also used as a basic tool for proving certain theorems in modern number theory. For example, it has been used to find multiplicative inverses in a finite field.

Propositions 30 and 32 of Book 7 together are equivalent to the fundamental theorem that every positive integer can be written as a product of primes in an essentially unique way. These propositions are given as follows:

Proposition 30 (of Book 7)

If a number is a product of two numbers and if a prime number is a factor of the product, then the prime number is also a factor of one of the original two numbers. ■

Proposition 32 (of Book 7)

Any number is either of a prime number or a multiple of prime numbers. ■

Book 8 deals with numbers in geometrical sequences as well as with number theory. Proposition 20 of Book 9 proves the infinitude of prime numbers. The construction of *perfect numbers* is given in Proposition 36 of Book 9.

Book 10 deals with *commensurable* numbers and *incommensurable* numbers. The definitions of *commensurable* and *incommensurable* are given at the beginning of Book 10. Commensurable numbers can be considered to be rational, whereas incommensurable numbers can be considered to be irrational. Books 11 through to 13 deal with solid geometry. Book 11 generalizes the results of Books 1 through 6 to solids (i.e., figures of three dimensions). Book 12 studies volumes of cones, pyramids, cylinders, and spheres. For example, Proposition 10 of Book 12 shows that the volume of a cone is one-third of the volume of the corresponding cylinder. *Elements* ends with Book 13, which discusses the properties of the five regular polyhedrons. Book 13 is largely based on an earlier treatise by an Athenian mathematician, Teaetetus, who first proved that there can be only five regular polyhedrons (i.e., regular tetrahedron, regular hexahedron, regular octahedron, regular dodecahedron, and regular icosahedron).

Theon of Alexandria (c. 335–c. 405 AD) was a Greco-Egyptian scholar and a mathematician in the fourth century. He edited and arranged Euclid's *Elements*. His edition was widely used and had become the only surviving Greek source until the 19th-century discovery of another source in the Vatican Library [10]. The Arabs received *Elements* from Byzantine around 760 AD. This version was translated into Arabic around 800 AD. Although some Western scholars probably knew that *Elements* existed in Byzantine, there is no existing record of Euclid's *Elements* having been translated into Latin before the 12th century. It was lost to West Europe until c. 1120, when an English monk translated it into Latin from the Arabic translation.

The first printed edition of Euclid's *Elements* appeared in 1492 in Venice. Since then it has been translated into many languages and published in many different editions. Theon's Greek edition was recovered in 1533. Some of the Greek old texts still survive and can be found in the Vatican Library and in the Bodleian Library, the main research library at the University of Oxford.

As a mathematical textbook, *Elements* is a masterpiece. It has been very influential in many areas of science. For example, Nicolaus Copernicus (1473–1543), Johannes Kepler (1571–1630), Galileo Galilei (1564–1642), and Isaac Newton (1642–1727) were all strongly influenced by its axiomatic deductions, logical approach, and rigorous proofs. At around age 40, Abraham Lincoln studied *Elements* for training in reasoning as a lawyer. His law partner Bill Herndon (1818–1891), the biographer of Abraham Lincoln, tells how late at night Lincoln would lie on the floor studying Euclidean geometry. Lincoln's logical speeches and some of his phrases, such as "dedicated to the proposition that all men are created equal" in the Gettysburg Address (November 19, 1863), are attributed to his reading of Euclid's *Elements* [4].

The following story by Bertrand Russell (1872–1970), a famous British philosopher, logician, and mathematician, is also well known. It is written in his autography: "At the age of eleven, I began 'Euclid', with my brother as my tutor. This was one of the great events of my life, as dazzling as first love. I had not imagined that there was anything so delicious in the world" [8].

REFERENCES

1. Euclid (author), T. L. Heath (trans.), and D. Densmore (ed.), *Euclid's Elements*, Green Lion Press, Santa Fe, NM, 2002.
2. Wikipedia, Euclid, http://en.wikipedia.org/wiki/Euclid.
3. Wikipedia, Euclidean Algorithm, http://en.wikipedia.org/wiki/Euclidean_algorithm.

4. Wikipedia, Euclid's Elements, http://en.wikipedia.org/wiki/Euclid's_Elements.
5. Wikipedia, Euclid of Alexandria, http://www-history.mcs.st-andrews.ac.uk/Biographics/Euclid.html.
6. D. A. Flower, *The Shores of Wisdom*, Pharos Publications, UK, 1999.
7. D. E. Knuth, *The Art of Computer Programming* (Vol. 2, 2nd ed.), Addison-Wesley, Reading, MA, 1981.
8. M. A. Plastow and Y. Igarashi, *The Mind of Science*, Kyoritsu-shuppan, Tokyo, 1989.
9. B. L. van der Waerden, *Geometry and Algebra in Ancient Civilizations*, Springer-Verlag, Berlin, 1983.
10. Wikipedia, Theon of Alexandria, http://en.wikipedia.org/wiki/Theon_of_Alexandria.

CHAPTER 6

Diophantus of Alexandria and *Arithmetica*

Diophantus of Alexandria was a Greek mathematician who lived in Alexandria, Egypt, probably from sometime between 200 and 214 AD to sometime between 284 and 298 AD. Diophantus's age of 84 years can be determined from the solution to a linear equation given in an inscription (a mathematical poem) on his tomb. The English translation is as follows:

> God vouchsafed that he should be a boy for sixth part of his life; when a twelfth was added, his cheeks acquired a beard; He kindled for him the light of marriage after a seventh, and in the fifth years after his marriage he granted him a son. Alas! Late-begotten and miserable child, when he had reached the measure of half his father's life, the chill grave took him. After consoling his grief by his science of numbers for four years, he reached the end of his life.

Diophantus is often referred to as the Father of Algebra, and is best known as the author of a series of books called *Arithmetica*. These books were a work on the solution of algebraic equations and on various aspects of number theory. However, there is little biographical information about Diophantus.

Arithmetica is the major work of Diophantus. It is a collection of problems on both determinate and indeterminate algebraic equations with their numerical solutions. Unfortunately, of the original 13 volumes, only 6 have survived: volumes I to III and volumes VIII to X of the original text.

Besides these Greek original volumes, four volumes of Arabic translations were discovered in the Astan Quds Library in Meshed, Iran, in 1970 [2]: volumes IV, V, VI, and VII of the original text [2]. These 10 volumes (6 Greek original volumes and 4 volumes of Arabic translations) contain about 200 problems with their numerical solutions. Diophantus considered only rational numbers in his books. Here, we list several examples of the problems from *Arithmetica*, where Problem I-1 means the first problem in volume I, Problem I-2 means the second problem of volume I, and so on:

Problem I-1

The sum of two numbers is 100, and the difference between these numbers is 40. Find these numbers.

SOLUTION

Let x be the smaller number. Then the larger number is $x + 40$. Hence, $2x + 40 = 100$. The required numbers are 30 and 70.

Problem I-2

The sum of two numbers is 60, and the ratio of the two numbers is 3:1. Find these numbers.

SOLUTION

Let the smaller one be x. Then the larger one is $3x$. Then $x + 3x = 60$. Hence, the required numbers are 15 and 45.

Problem I-27

Find two numbers such that their sum and product are given numbers.

Necessary condition: The square of half the sum must exceed the product by a square number.

Suppose that the sum and product of two numbers are 20 and 96, respectively. Find the two numbers.

SOLUTION

Let $2x$ be the difference of the required numbers. Then the two numbers are $10 - x$ and $10 + x$. Hence, $(10 + x)(10 - x) = 100 - x^2 = 96$. Therefore, $x = 2$, and the required numbers are 8 and 12.

(Note that given two numbers a and b, $((a + b)/2)2 - ab = ((a - b)/2)2$. The necessary condition of this problem is derived from this equation.)

Problem II-6

Find two numbers having a given difference and a number such that the difference of their squares exceeds their difference by a given number.

Necessary condition: The square of their difference must be less than the sum of the said difference and the given excess of the difference of the squares over the difference of the numbers.

Let the two numbers be x and y. Suppose that the given difference is $x - y = 2$ and the given excess $x^2 - y^2 - (x - y) = 20$. Find the two numbers x and y.

SOLUTION

The necessary condition is symbolically $(x - y)^2 < (x - y) + x^2 - y^2 - (x - y) = x^2 - y^2$. Since $x - y > 0$, the necessary condition means $x > y > 0$.

Then $4y + 4 = 22$. It follows that the required numbers are $y = 9/2$ and $x = 13/2$.

Problem II-8

Divide a given square number into two square numbers. Let the given square number be 16.

SOLUTION

Let x^2 be one of the required squares. Then, $16 - x^2$ must be equal to a square. Take a square of the form $(mx - 4)^2$, where m is any integer and 4 is from the square root of 16. For example, take $(2x - 4)^2$, and equate it to $16 - x^2$. Hence, $4x^2 - 16x + 16 = 16 - x^2$, or $5x^2 = 16x$, and then $x = 16/5$. Therefore, the required squares are 256/25 and 144/25.

Problem IV-3

Find two square numbers such that their sum is a cubic number.

SOLUTION

Let x^2 and $4x^2$ be a smaller square number and a larger square number, respectively. Let $x^2 + 4x^2 = x^3$. Then $x = 5$. Therefore, 25 and 100 are two square numbers satisfying the condition of the problem. We can find other pairs of such square numbers in a similar way. For example, let such a pair of square numbers be x^2 and $9x^2$, and let $x^2 + 9x^2 = x^3$. Then $x = 10$. Therefore, 100 and 900 are also a pair of two square numbers satisfying the condition.

Problem V-7

Find two numbers such that the sum of the numbers is 20 and the sum of the cube of the first and the cube of the second is 2240 (i.e., $x + y = 20$ and $x^3 + y^3 = 2240$).

SOLUTION

Let $x = 20 \times 1/2 + s$ and $y = 20 \times 1/2 - s$. Then $x^3 + y^3 = (x + y)((x + y)^2 - 3xy)$ $= 20(400 - 3(100 - s^2)) = 20(100 + 3s^2) = 2240$. Hence, $100 + 3s^2 = 112$ and $s = 2$. Therefore, $x = 12$ and $y = 8$.

Problem VIII-15

Find three numbers such that the sum of any two multiplied by the other is a given number. Let (first + second) × third = 35, (second + third) × first = 27, and (third + first) × second = 32.

SOLUTION

Let the third be x. Then (first + second) = $35/x$. Assume first = $10/x$ and second = $25/x$. Then we have

$$250/x^2 + 10 = 27 \text{ and } 250/x^2 + 25 = 32$$

These equations are inconsistent, but they would not be if $25 - 10$ were equal to $32 - 27 = 5$. From this observation we have to divide 35 into two parts, replacing 25 and 10, such that their difference is 5. The parts are 15 and 20. Therefore, we may take $15/x$ as the first number, $20/x$ as the second, and we now have

$$300/x^2 + 15 = 27 \text{ and } 300/x^2 + 20 = 32$$

Then the third = $x = 5$, the first = 3, and the second = 4.

A Diophantine equation is an indeterminate polynomial equation that allows integer solutions only. Diophantus made a study of such equations in his books. He was one of the first mathematicians who used symbolism in algebra. The following are some examples of Diophantine equations, where x, y, and z are unknown variables, and n and k are given integers:

1. $ax + by = k$: A linear Diophantine equation.

2. $x^n + y^n = z^n$: For $n = 2$, there are infinitely many solutions for x, y, and z. The integer solutions are called *Pythagorean triples*.

3. $x^2 - ny^2 = +1$ or -1: Pell's equation. It is named after English mathematician John Pell (1611–1685). It was also studied by Indian mathematician Brahmagupta (598–c. 668) in the 7th century as well as French mathematician Pierre de Fermat (1601–1665) in the 17th century.

4. $4xyz = n\,(xy + xz + yz)$, equivalently $4/n = 1/x + 1/y + 1/z$: The Erdös-Straus conjecture states that for every integer $n \geq 2$, there exists a solution in x, y, z, all as positive integers. Paul Erdös (1913–1996) and Ernst G. Straus (1922–1983) formulated the conjecture in 1948. Computer searches have verified the truth of the conjecture up to $n \leq 10^{14}$, but proving it for all integers n remains an open problem.

Diophantus wrote several other books in addition to *Arithmetica*, but very few of them have survived. Diophantus referred to his book called *The Porisms*, but this book has not been found. It is not known whether *The Porisms* is one of the lost volumes of *Arithmetica*.

A *polygonal number* is a number represented as dots or pebbles arranged in the shape of a regular polygon. For example, *triangular numbers* are 1, 3, 6, 10, 15, …, and *square numbers* are 1, 4, 9, 16, 25, …, and *pentagonal numbers* are 1, 5, 12, 22, 35, …. These are examples of polygonal numbers. Diophantus is also known to have written on polygonal numbers.

In Western Europe, Diophantus was forgotten until the 15th century, though a portion of *Arithmetica*, like other ancient Greek texts, was known to some medieval Byzantine scholars and Arabic scholars. In 1463 a German mathematician, Johannes Müller von Königsberg (1436–1476), wrote that no one had yet translated the 13 volumes of *Arithmetica* from Greek into Latin. In 1570 an Italian mathematician, Rafael Bombelli (1526–1572), translated a portion of the original Greek text of *Arithmetica*, but it was never

Example 6.1

Find a solution for the following linear Diophantine equation:

$$2072x + 1665y = 37$$

By the Euclidean algorithm, we have the following sequence of divisions:

$$2072 = 1 \times 1665 + 407,$$

$$1665 = 4 \times 407 + 37,$$

$$407 = 11 \times 37 + 0.$$

Hence, we find $gcd\,(2072, 1665) = 37$. Tracing backward the second and first equations above, we obtain the following equations:

$$37 = 1665 - 4 \times 407$$

$$= 1665 - 4 \times (2072 - 1665)$$

$$= -4 \times 2072 + 5 \times 1665.$$

Thus, we find $x = -4, y = 5$ to the given linear Diophantine equation.

More generally, $x = -4 - 1665t, y = 5 + 2072t$ is also a solution to the linear Diophantine equation for any integer t.

published. The best-known Latin translation of *Arithmetica* was made by the French mathematician Claude-Gospar Bachet de Meziriac (1581–1638) in 1621. That translation was widely available in Western Europe.

Pierre de Fermat, a French lawyer and mathematician, owned a copy of the 1621 edition of the Latin translation of *Arithmetica*. Around 1637 he wrote a memo, so called *Fermat's last theorem*, in the margin of his copy as follows:

If an integer n is greater than 2, then $x^n + y^n = z^n$ has no solutions in non-zero integers x, y, and z. I have a truly marvelous proof of the proposition, but this margin is too narrow to contain the proof.

Fermat's claimed proof for his proposition was never found. It is believed that Fermat did not have the proof. The problem of finding a proof for Fermat's last theorem had been an unsolved problem for more than 350 years. A proof for the theorem was finally found in 1994 by the British mathematician Andrew John Wiles (1953–) after working on the problem for 7 years.

In mathematics, the modularity theorem (also called the Taniyama–Shimura–Weil conjecture) states that elliptic curves over the field of rational numbers are related to modular forms. Wiles proved that the modularity theorem for semistable elliptic curves was sufficient to imply Fermat's last theorem. He realized that a proof of a limited form of the modularity theorem might be in reach. He decided to devote all of his research time to this problem. In 1993, he presented his proof to the public for the first time at a conference. However, it turned out that the proof contained a fundamental gap. The crucial idea for circumventing the gap came to him in 1994. Together with his former student Richard Taylor (1962–), he published a second paper that circumvented the gap and completed the proof of Fermat's last theorem. Both the first paper by Wiles and the second paper by Taylor and Wiles appeared in 1995 in the *Annals of Mathematics* published by Princeton University [7, 8].

Hilbert's problems form a list of 23 problems in mathematics published by the German mathematician David Hilbert (1862–1943) in 1900. He presented 10 problems among them at the Paris conference of International Congress of Mathematicians. Hilbert's 10th problem asked to find an algorithm for determining whether an arbitrary Diophantine equation has a solution. The Russian mathematician Yuri Vladimirovich Matiyasevich (1947–) proved that no such algorithm is possible in 1970. That is, he proved the impossibility of obtaining a general algorithm for Hilbert's 10th problem, making it unsolvable (see Chapter 16).

Diophantus's *Arithmetica* has been, over the centuries, the source for many algebraic theorems and has influenced significantly the development of number theory, mathematical notation, and the use of symbolism in algebra.

REFERENCES

1. T. L. Heath, *Diophantus of Alexandria*, Cambridge University Press, Cambridge, UK, 1910.
2. N. Schappacher, *Diophantus of Alexandria*: A Text and Its History, manuscript, 2005.

3. Wikipedia, Diophantus, http://en.wikipedia.org/wiki/Diophantus.
4. B. L. van der Waerden, *Geometry and Algebra in Ancient Civilization*, Springer-Verlag, Berlin, 1983.
5. Wikipedia, Fermat's Last Theorem, http://en.wikipedia.org/wiki/Ferma's_Last_Theorem.
6. Wikipedia, Diophantus Equation, http://en.wikipedia.org/wiki/Diophantine_equation.
7. A. Wiles, Modular Elliptic Curves and Fermat's Last Theorem, *Annals of Mathematics*, 141(3), 443–551, 1995.
8. R. Taylor and A. Wiles, Ring-Theoretic Properties of Certain Hecke Algebras, *Annals of Mathematics*, 141(3), 553–572, 1995.

CHAPTER 7

Secret Writing in Ancient Civilization

7.1 STEGANOGRAPHY

It is difficult to specify when secret writing started in ancient civilization. Some of the earliest accounts of secret writing date back to Herodotus (c. 484–c. 425 BC), who in his text *Histories* describes the conflict between Greece and Persia in the fifth century BC. According to *Histories*, secret writing by a Spartan saved Greece from being conquered by Xerxes I (the Great) of Persia, who reigned from 485 to 465 BC. Herodotus was an ancient Greek historian and was regarded as the Father of History in Western culture. The Greek word *historia* passed into Latin and took on its modern meaning of history. *Histories* was divided into nine books by Alexandrian editors and was structured as a dynastic history of ancient Persian kings. The following story was recorded in Book 7 of *Histories*:

> Xerxes the Great spent about 5 years secretly assembling the greatest fighting force to launch an attack on Greece. However, the Persian military buildup was discovered by Demaratus when he lived in a Persian city, Susa, after having been expelled from Sparta. Before his exile, he had been the king of Sparta from 515 to 491 BC. Demaratus wanted to send a secret message to warn the Spartans of Xerxes's invasion plan to Greece. He wrote his secret message on a wooden tablet, and then covered it over with wax. When the

wooden tablet reached its destination, the Spartans scraped off the wax and found the message written on the wood underneath. As a result of this warning, Greeks began to construct 200 warships.

On September 23, 480 BC, when the Persian fleet approached the Bay of Salamis near Athens, they found the Greek navy prepared for battle. Within a day, the Persian fleet was defeated by the Greeks.

Secret communication achieved by hiding the existence of a message is known as *steganography*. Demaratus's strategy for secret communication is an example of steganography. It is the combination of Greek origin words, *steganos* meaning "covered or protected" and *graphein* meaning "to write."

The first recorded use of the term *steganography* was in 1499 by a German scholar, Johannes Trithemius (1462–1516), in his book *Steganographia* (published in Frankfurt). Another example of ancient steganography described by Herodotus in *Histories* is a tattooed message on the shaved head of a trusted slave. After his hair had grown, the slave was dispatched to the desired destination with the message hidden in his hair (i.e., at the time of dispatch it is unreadable).

In ancient China, secret messages were often written on fine silk. They were first scrunched into tiny balls and then covered by a layer of wax. The messenger would then swallow the waxed ball and carry it in his stomach to its destination.

7.2 CRYPTOGRAPHY

Steganography suffers from a serious weakness. If the messenger is searched and the hidden message is discovered, its contents are revealed at once. Another strategy to hide a message is via *cryptography*, which had also been developed in ancient civilization. The term *cryptography* originates from the combination of Greek origin words *krypts*, meaning "hidden," and *graphein*, meaning "writing." Cryptography hides the meaning of a message by a process known as encryption [2, 4]. It is not an attempt to hide the existence of the message. Cryptography was concerned with message confidentiality by converting the message from a comprehensible form into an incomprehensible one so that the interceptors or eavesdroppers could not understand the encrypted message without secret knowledge of how to decrypt it.

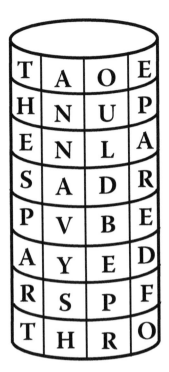

FIGURE 7.1   An example of a message written on the scytale.

In cryptography, a secret protocol called a secret key is agreed to beforehand between the sender and a legitimate receiver. The receiver can reverse the converted message by using the secret key to make it comprehensible. Cryptosystems can be classified into systems of *transposition* and *substitution*.

In transposition, the letters of a message are rearranged. The specific arrangement protocol of letters needs an agreement between the sender and the receiver beforehand, but the agreement is kept secret from the enemy. A typical example of transposition can be seen in the Spartan military cryptographic device called *scytale*, dating back to the fifth century BC. The scytale is a wooden cudgel around which a strip of parchment is wound, as shown in Figure 7.1. The sender writes the message along the length of the scytale, and then unwinds the strip, which now looks like a sequence of meaningless letters. To recover the message, the receiver wraps the parchment strip around a scytale of the same diameter as used by the sender. By doing so, the receiver can read the original message along the length of the scytale.

As an example, we show the message "The Spartan navy should be prepared for the attack by the Persian fleet" written on a parchment strip that is wound around a scytale (Figure 7.1). Although we have spaces between words and some punctuation marks in plaintexts, these are usually omitted in cryptotexts.

For very short messages such as single words, transposition cryptography is relatively insecure, because there are only a limited number of ways for rearranging a small number of letters. However, as the number of letters in messages increases, the number of possible arrangements rapidly explodes. Therefore, for a relatively long message, it is very hard to restore the original message (i.e., decipher it) unless the arranging process is known.

The alternative to transposition in cryptography is substitution. In substitution, the plaintext letters are replaced with substitutes. The substitutes are kept in the cryptotext in the same order as their originals in the plaintext. If the use of substitutes remains unaltered throughout the text, the cryptosystem is called monoalphabetic. As far as we know, all of the ancient substitution systems were monoalphabetic.

Caesar cipher is the most widely known among monoalphabetic substitution cryptosystems. It was named after Julius Caesar (100–44 BC), a Roman military and political leader. He played a critical role in the transformation of the Roman Republic into the Roman Empire. Caesar was also considered during his lifetime to be one of the best orators and authors in Latin. *Commentarii de Bello Gallico* (Commentaries on the Gallic War) was a series of books describing military campaigns waged by Julius Caesar against several Gallic tribes. It consists of eight volumes; volumes 1 to 7 were written by Caesar himself, while volume 8 was written by one of his subordinates.

Caesar cipher is based on substitutions in the following way. Each letter in the cryptotext is obtained from its corresponding letter in the plaintext by advancing k steps in the alphabet. At the end of the alphabet one goes cyclically to the beginning. Caesar cipher is also called *shift cipher* [1, 3]. Thus, for $k = 3$, substitutions are as follows:

Plaintext: A B C D E F G H I J K L M N O P Q R S T U V W X Y Z

Cryptotext: D E F G H I J K L M N O P Q R S T U V W X Y Z A B C

In this cryptosystem, the plaintext WE GOT A GREAT VICTORY is encrypted as ZH JRW D JUHDW YLFWRUB.

Julius Caesar described in his *Commentarii de Bello Gallico* how he sent an encrypted message to Marcus Tullius Cicero (106–43 BC), a Roman philosopher, statesman, and lawyer. It was the substitution system in monoalphabetic, but did not employ the Caesar cipher. The Latin letters in the plaintext were replaced by Greek ones in the cryptotext to Cicero. The historical record that Julius Caesar actually used the Caesar cipher comes from the writings of Gaius Suetonius Tranquillus (c. 70–140 AD), a historian during the Roman Empire. According to Suetonius, the shift in the alphabet in the Caesar cipher was three, as shown above.

In Caesar cipher and other similar cryptosystems, the following numerical encoding is convenient:

A	B	C	D	E	F	G	H	I	J	K	L	M
0	1	2	3	4	5	6	7	8	9	10	11	12
N	O	P	Q	R	S	T	U	V	W	X	Y	Z
13	14	15	16	17	18	19	20	21	22	23	24	25

If we use this numerical encoding, the encryption and decryption process in Caesar cipher can be expressed via modular arithmetic. Thus, according to Caesar cipher, each letter α in a plaintext is encrypted as $\alpha + k$ modulo 26, and each letter β in a cryptotext is decrypted as $\beta - k$ modulo 26. For example, suppose $k = 3$, then *boy* is encoded to 1 14 24, and encrypted to 4 17 1. The cryptotext 4 17 1 is decrypted to 1, 14, 24. Note that $24 + 3 \equiv 24 + 3 - 26 \equiv 1 \bmod 26$, and that $1 - 3 \equiv 1 - 3 + 26 \equiv 24 \bmod 26$. Additional applications of modular arithmetic will be described more in Chapter 20.

The number of all possible keys for Caesar cipher is only 26, and the alphabetical order remains as in the sequence of substituted letters. From the security point of view, these are great disadvantages of Caesar cipher. Adversaries can easily break Caesar cipher by attempting to break the cryptotext for each possible key value of 0 to 25.

Caesar cipher is not the oldest monoalphabetic substitution cryptosystem. A system devised by Polybius (c. 203–120 BC) can be also considered a monoalphabetic substitution cipher. Polybius was a Greek historian of the Hellenistic period. He died about 30 years before the birth of Caesar. We explain the Polybius system using the Roman alphabet. Consider the following square matrix, nowadays called the *Polybius checkerboard* (also known as Polybius square) (Table 7.1).

TABLE 7.1   The Polybius Checkerboard

	A	B	C	D	E
A	A	B	C	D	E
B	F	G	H	I/J	K
C	L	M	N	O	P
D	Q	R	S	T	U
E	V	W	X	Y	Z

Each letter α will be represented by the pair of letters indicating the row and column in which α lies in the matrix above. For example, the representations of A, B, M, N, Y, and Z are AA, AB, CB, CC, ED, and EE, respectively. The plaintext "Let us build ships" is encrypted to CAAEDDDEDCABDEBDCAADDCBCBDCEDC. The Roman alphabet version of the Polybius system is monoalphabetic substitutions into the target alphabet {AA, AB, AC, AD, AE, BA, BB, …, EE}.

One of the earliest encryptions by monoalphabetic substitutions appears in the *Kama Sutra*, a group of texts written in the fourth century in India. Mallanaga Vatsyayana was a Hindu philosopher in the Vedic tradition, and is believed to have lived during the Gupta Empire (fourth to sixth centuries). His name appears as the author of the *Kama Sutra*, but its original manuscripts were dated back as far as the fourth century BC. The *Kama Sutra* recommends that women should study 64 arts, such as cooking, dressing, and preparing perfumes. Secret writing is one of the 64 arts, and it helped women conceal their affairs and liaisons. The key of the secret writing is to pair letters of the alphabet, and the encryption is to substitute each letter in a plaintext with its partner letter. We explain this principle using the Roman alphabet. Suppose that the pairing is (A, O), (B, S), (C, T), (D, U), (E, V), (F, Y), (G, W), (H, Z), (I, K), (J, N), (L, X), (M, P), (Q, R). Then LOTS OF LOVES is encrypted to XACBAYXAEVB.

Random ways of pairing the letters in an alphabet seem to offer a high level of security, because the number of possible pairings is very large. For the Roman alphabet, the number of possible pairing ways is $25 \times 23 \times 21 \times 19 \times 17 \times 15 \times 13 \times 11 \times 9 \times 7 \times 5 \times 3$, which is approximately 7.09×10^{12}. Of course, both the sender and the receiver must agree on a specific letter pairing arrangement in order for the system to be useful.

Although cryptography and steganography are independent, it is possible to combine these methods to increase security.

REFERENCES

1. Arto Salomaa, *Public-Key Cryptography*, Springer-Verlag, Berlin, 1990.
2. Simon Singh, *The Code Book: The Science of Secrecy from Ancient Egypt to Quantum Cryptography*, Anchor Books, New York, 1999.
3. Douglas R. Stinson, *Cryptography: Theory and Practice*, CRC Press, New York, 1995.
4. Wikipedia, Cryptography, http://en.wikipedia.or/wiki/Cryptography.

CHAPTER 8

The Abacus

8.1 THE EARLIEST ABACI

Abacus (plural: *abaci* or *abacuses*) is a Latin word that came from the Greek word *abax* or *abakon*, meaning "table" or "tablet," and possibly from the Hebrew word *abaq*, meaning "dust" or "sand." The abacus in its various forms is a calculating tool invented by ancient civilizations. The simplest of these tools, called a counting board or abacus, may have initially been made with lines drawn in the sand and pebbles placed between or on the lines as counters.

If we search the archeological record, however, we can find direct written evidence and clay artifacts related to the earliest known abaci from Mesopotamia. Sumer, an ancient civilization in the southern part of Mesopotamia, was established by proto-Euphratean people around 5000 BC. The Sumerians at first used tokens (clay objects of various shapes and sizes marked with symbols) to represent items for trade and accounting. By the third millennium BC, these tokens had become "calculi" or clay tokens used for arithmetic calculation, the results of which were written down on clay tablets in the form of abstract numbers. At this time, although the Sumerians had abstract numbers, they did not perform calculations directly with them; numbers were used only for record keeping.

For arithmetic calculations, the Sumerians used calculi, or clay tokens whose size or shape reflected the order of magnitude in their number system. The calculi were manipulated, depending on the arithmetic operation, by interchanging multiple smaller-level calculi for larger ones, or vice versa, until the solution was obtained. At some point, the Sumerians discovered that arranging the calculi on a table delineated with columns allowed the same sort of operations to be done more efficiently and could also be done using fewer uniformly shaped tokens [2].

It is estimated that the first abacus appeared in Sumer sometime between 2700 and 2300 BC [1]. At that time, the Sumerians had already developed a positional base-60 number system. The earliest abacus was a wooden or clay board, divided into columns labeled with the orders of magnitude of a base-60 number system. The various shaped calculi were replaced with uniform-sized tokens of clay or short reed sticks. Sumerians performed calculation with these tools, while written numbers were used mainly to record the results [2].

Mesopotamia became united under the rule of the Assyro-Babylonians around 2000 BC. These Semitic people adapted the Sumerians' cuneiform writing in their own language. Since they had a decimal number system, it was necessary to create conversion tables for the abaci and translate the base-60 results into decimal numbers [3]. Eventually, the abacus itself was converted to a decimal format. By the end of the third millennium BC, a radical transformation in calculation took place [1]. Rather than the use of tokens on the counting boards or abaci, the Babylonians began writing numbers directly into clay tablets, erasing and rewriting the numbers as each operation was carried out.

The Babylonians used their mathematics for complex astronomical calculations, and in the management of a large administrative bureaucracy supported by taxation and trade. The abacus, evolving over time, was no doubt used by both Babylonian scholars and others as a convenient tool for calculation.

Eventually, for portability and convenience, grooved wooden boards were invented. Later, more permanent tablets of stone (e.g., marble) and metal tablets appeared. People used the abacus to count numbers, perform arithmetic operations, and record calculated results. Traders and merchants needed a tool for both counting the number of goods bought and sold and calculating the price of their wares [4].

While the Babylonians enjoyed an abundance of agricultural products thanks to the fertile farmland of Mesopotamia, they lacked the natural resources necessary to maintain their growing city-state civilization. For this reason, they developed vast trade networks extending to India, Persia, and many cultures and states surrounding the Mediterranean Sea. It is thought that the Babylonian abacus was introduced throughout the Middle East and the Mediterranean area via trade and commercial networks. As it was adopted by various cultures, the design of the abacus was modified to suit their social needs and reflect the appropriate language and number system [11].

8.2 THE SALAMIS TABLET AND THE ROMAN HAND ABACUS

The oldest surviving counting board or abacus is the *Salamis Tablet* used around 300 BC, which was discovered in 1846 on Salamis Island, an ancient Greek city-state off the coast of Cyprus. It is a slab of white marble with some marks. In the upper center of the tablet there is a set of five parallel lines equally divided by a vertical line. Below these lines is a wide space with another set of 11 parallel lines divided by a vertical line. Three sets of Greek symbols (numbers from the *acrophonic* system, or the first Greek number system) are arranged along the left, right, and bottom edges of the tablet. The slab is 149 cm long, 75 cm wide, and 4.5 cm thick. It was used by the Greeks, but it is thought that its design was based on the Babylonian counting board [7, 10]. The following are some examples of numeral representations in acrophonic:

TABLE 8.1   Examples of Acrophonic Numerals

I	II	III	IIII	Γ	ΓI	ΓII	ΓIII	ΓIIII
1	2	3	4	5	6	7	8	9
Δ	H	X						
10	100	1000						

As with the Salamis Tablet, pebbles are used to represent numbers. Numbers between 0 and 4 were generally represented by a number of pebbles. A system of lines on the slab serves to give weights to numbers by powers of 10. A pebble between the lines represented the number 5. Pebbles on the right side of the vertical line represent positive digits, and those on the left side represent negative digits. For example, the number 4 might be represented as a pebble above the right side of the first line plus a pebble on the left side of the first line, which represents 5 − 1 = 4. These two pebbles represent the same number as the representation of four pebbles on the right side of the first line. Likewise, the number 90 might be represented as a pebble on the right side of the third line plus a pebble of the left side of the second line. It was possible to perform additions and subtractions of large numbers [5, 7, 13]. The Salamis Tablet is currently at the National Museum of Epigraphy in Athens, Greece.

Once the practical counting board had found its way to the Mediterranean region, it was adopted by the ancient Greeks, Romans, and Egyptians. In addition to the Salamis Tablet, concrete evidence of the use of abaci was found on the Darius Vase (c. fourth century BC), discovered in an ancient burial site in Canossa, Italy. One scene on the vase depicts a seated royal treasurer using an abacus to calculate as a man in front of him is presenting more tribute to be counted. Like the Salamis Tablet, the abacus on the Darius Vase is marked with Greek symbols for drachmas in decimal orders of magnitude. This beautiful vase is in the Museo Nationale in Naples, Italy [16]. The Etruscan Cameo (fifth century BC), a 1.5 cm high carved artifact from ancient Etruria, is another object clearly showing the use of a counting board. Here again we see an abacist sitting at a table on which tokens have been placed. He has a tablet in his left hand on which he has recorded his results in Etruscan numerals. This cameo is in the Cabinet des Medailles, Paris [17].

The ancient Romans designed an early portable calculating device called the *Roman hand abacus* for use by merchants, bankers, engineers, architects, tax collectors, and others. It is a base-10 version of the previous Babylonian abacus. There are interesting similarities between the Roman hand abacus and the Salamis Tablet. The lower slots and upper slots on the Roman hand abacus are presumably the proper promotion factors of lines and spaces between lines on the Salamis Tablet. A Roman hand abacus, which is currently on display at the Bibliotheque Nationale de France, in Paris, was made in the first century AD. Replicas of similar Roman hand abaci can be found in the Science Museum and in the British Museum, both in London. It greatly reduced the time for performing the basic arithmetic operations, as opposed to hand calculating with Roman numerals [6].

The Roman hand abacus consists of a metal plate with nine parallel columns of slots. The first two columns on the right side of the hand abacus are for calculating fractions. The remaining seven slots for calculating integers are divided into an upper and a lower deck. Each of the slots on the upper deck has one sliding bead, and each of the slots on the lower deck has four. Each bead in the lower deck slots represents a unit of the power of 10. The user of the Roman abacus slides the lower beads up toward the center for numbers less than 5 (50, 500, etc., depending on the column). To represent 5 units, one upper bead is slid down to the center. For example, the number 6 is one bead from the upper deck and one bead from the lower deck, both moved to the center, while the number 7 is one upper bead and two lower beads in the center [6].

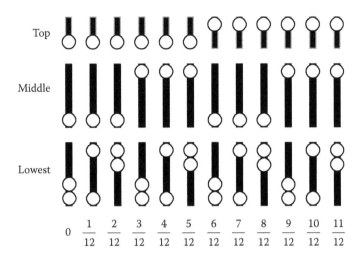

FIGURE 8.1   Representations from 0 to 11/12 on the rightmost slot.

The two right fraction slots on the Roman hand abacus are base-12. This is because the Roman *as* or pound was divided into 12 *uncial* or ounces. Fractions of Roman currency were expressed in terms of pounds and ounces. Thus, for fractional accounting purposes, the base-12 parts of the abacus were extremely convenient. Both of the two right columns (slots) for calculating fractions were used to count from 0 to 11/12 in different ways (See Figure 8.1). The second column from the rightmost is divided into the lower slot and the upper slot: five beads are in the lower slot and one bead is in the upper slot. The column counts up to 5 ounces (5/12 pound) in the lower slot and carries into the upper slot on a count of 6 ounces (6/12 pound), repeats to a count of 11 ounces (11/12 pound), and then carries into the decimal units (pounds) on a count of 12 ounces (1 pound). Thus, the column can count from 0 to 11/12 pound. In other words, the bead in the upper deck represents 1/2 pound and each bead in the lower slot represents 1/12 pound. On the other hand, the rightmost column is divided into three sections: the lowest slot, the middle slot, and the top slot have two beads, one bead, and one bead, respectively. The rightmost column can count from 0 to 11/12 ounce, and then carries into the next right column into 1 ounce (unit) of a count of 12/12 ounce [6]. In other words, the bead in the top slot represents 1/2 ounce, the bead in the middle slot represents 1/4 ounce, and either of two beads in the lowest slot represents 1/12 ounce. Each of 12 representations (from 0 to 11/12 ounce) on the rightmost column is displayed in Figure 8.1.

8.3 THE CHINESE ABACUS

While it is clear that the Sumerian abacus underwent many adaptations and had spread throughout the lands of the Middle East and Mediterranean, there is also evidence of the use of abaci from ancient times in Asian civilizations. In the second century BC, the Chinese calculated with small bamboo sticks or rods on a board. They used a positional decimal system, and like the early Sumerians, while they had written numbers, they did not calculate with them directly [2]. The earliest known written documentation about the Chinese abacus is probably in a book (c. 190 AD) on the Eastern Han dynasty (25–225 AD). However, the exact design of the earliest Chinese abacus is not known.

Similarly, between the second century BC and the third century AD, Indian mathematicians computed on a sand-covered board, lined with columns. The Indian word *dhuli-karma*, or "sand work," means higher computations. The method of calculation, however, was discontinued between the fourth and sixth centuries AD when the use of a written decimal number system spread. It is this transformation that is also considered to be the start of modern written arithmetic [1].

Several centuries later, there is further evidence of more advanced abaci in use in China. The panoramic painting *Along the River during the Qingming Festival* by the Song dynasty artist Zhang Zeduan (1085–1145 AD) depicts the daily life of people and the landscape of the capital Bianjing (today's Kaifeng). In the scene, a Chinese abacus can be seen beside an account book. Furthermore, two books demonstrating the use of the Chinese abacus, or *suanpan*, appeared in the Shun-hi dynasty (1170–1190 AD). These books are the *Pan chu tsih* and the *Tseu pan tsih*, where *pan*, *chu*, *tseu*, and *tsih* mean "counting board," "counting beads," "operations," and "book," respectively. Based on these and other sources, it seems the suanpan is a relatively late development of the abacus in China, appearing in the 11th or 12th century. No definite description of the abacus in China before the 11th century has been found [14].

The number of rods in the traditional suanpan is usually odd, and never less than nine. There are five beads (called *earth* beads) on each rod in the lower deck and two beads (called *heaven* beads) on each rod in the upper deck. These beads are mounted on rods and slide up and down within decks. This style suanpan is also referred as a 5-2 suanpan. The 5-2 suanpan appeared in China around 1200 AD. The beads are rounded and made of wood, stone, or ivory. The rods are made of bamboo, wood, ivory, or metal.

Suanpan beads are counted by moving them up or down toward the crossbeam. The upward move of one bead in the lower deck means an increase by one place-value, while the downward move of one bead in the upper deck means an increase by five place-values. The beads can be quickly reset to the starting position by a quick jerk along the horizontal axis to move all the beads away from the horizontal beam separating the upper deck and the lower deck [8].

Any number between 0 and 9 can be displayed on each rod without using the uppermost bead in the upper deck and the lowest bead in the lower deck. Using all the beads on each rod, any number between 0 and 15 can be displayed on the rod. Thus, the 5-2 suanpan can be used for both decimal and hexadecimal computation. None of the beads is redundant for hexadecimal calculation with the 5-2 suanpan. However, the uppermost beads in the upper deck and the lowest beads in the lower deck are redundant for decimal calculation with the 5-2 suanpan. The traditional Chinese system of weights was hexadecimal. For example, 1 *jin* (600 grams) equals 16 *liand* (37.5 grams). The computations in decimal and hexadecimal were quite similar except for the usage of the uppermost beads in the upper deck and the lowest beads in the lower deck [15].

Later the 5-2 suanpan underwent modification, and appeared as a 5-1 suanpan in the Ming dynasty (1336–1644 AD). It is illustrated in a Chinese book from the 16th century [8] and has one bead in the upper deck and five beads in the lower deck on each rod. The 5-1 suanpan was mainly used for decimal calculations, while the 5-2 suanpan survived until the 19th century. Various efficient suanpan techniques have been developed for addition, subtraction, multiplication, division, and square and cubic roots in China. Multiplication required the use of the multiplication table, and division required the division table. The users of the suanpan usually learned these tables by heart.

8.4 THE JAPANESE ABACUS

The use of the suanpan spread from China to Korea around 1400 AD, where it is called *jupan, jusan,* or *supan.* During the 15th century, the 5-2 suanpan found its way also to Japan via the Korean peninsula, where it was called *soroban.* The Chinese division table was also introduced at the same time. The Chinese division table was called *hassan,* meaning "eight classes of division calculations." The method of using the division table was called *kyukihou* ("nine" returning method). Kanbei Mouri (also known as Shigeyoshi Mouri) was an influential Japanese mathematician during the

Azuchi-Momoyama era (1573–1603) and early Edo era (1603–1868). He instituted a mathematics school in Kyoto in the 16th century, spreading the kyukihou (i.e., the division method using the hassan). Mitsuyoshi Yoshida (1598–1672), one of his students, was the author of *Jinkouki* (A Book of Very Large and Very Small Numbers). This book is an old extant Japanese textbook on mathematics containing arithmetic techniques of the soroban. The book was widely read in the Edo and Meiji eras (Edo, 1603–1868; Meiji, 1868–1912). *Jinkouki* contributed greatly to spreading the soroban in Japan [9].

The 5-1 soroban appeared in Japan around 1850 at a time when the basis for Japanese currency shifted from hexadecimal to decimal [9]. The division method using the division table was rather complicated. It was replaced by a division method using the standard multiplication table around 1935, soon after the 4-1 soroban appeared. The 4-1 soroban has one bead in the upper deck and four beads in the lower deck on each rod. It became the standard configuration of the soroban in Japan.

In 1946, an exciting contest took place between the Japanese soroban champion at that time and an American expert of the electronic calculator in Tokyo, under the sponsorship of the U.S. Army newspaper, *The Stars and Stripes*. The newspaper reported that the soroban champion defeated the expert of the electronic calculator at additions, subtractions, divisions, and mixed operations. The expert of the electronic calculator defeated the soroban champion only at multiplications [12].

The soroban remained in use in Japan in commercial businesses until the 1960s, long after the abacus had been abandoned elsewhere. Even in banks and the accounting departments of major companies, calculations were performed on the soroban until the 1950s. Although the soroban had faded from practical use after the appearance of portable electronic calculators, its use continues to be taught in schools in Japan. The soroban is still manufactured in Japan today, though pocket electronic calculators have proliferated and are much more convenient to use.

REFERENCES

1. G. Ifrah, *The Universal History of Computing*, John Wiley & Sons, New York, 2001.
2. G. Ifrah, *The Universal History of Numbers*, John Wiley & Sons, New York, 2000.
3. C. Clawson, *The Mathematical Traveler: Exploring the Grand History of Numbers*, Perseus Publishing, New York, 1994.
4. A Brief History of the Abacus, http://www.ee.ryerson.ca/~elf/abacus/history.html.

5. Wikipedia, Abacus, http://en.wikipedia.org/wiki/Abacus.
6. Wikipedia, Roman Abacus, http://en.wikipedia.org/wiki/Roman_abacus.
7. Wikipedia, Salamis Tablet, http://en.wikipedia.org/wiki/Salami.
8. Wikipedia, Suanpan, http://en.wikipedia.org/wiki/Suanpan.
9. Wikipedia, Soroban, http://en.wikipedia.org/wiki/Soroban.
10. Wikipedia, An Overview of Babylonian Mathematics, http://www.history. mcs.st.andrews.ac.uk/HistTopics/Babylonian_mathematics.html.
11. K. E. Carr, West Asia Mathematics, Kidipede: History for Kids, 2012, http:// www.historyforkids.org/leam/westasia/science/math.htm
12. T. Kojima, *The Japanese Abacus: Its Use and Theory*, Charles E. Tuttle, Tokyo, 1954.
13. S. K. Stephenson, Roman Hand Abacus Is Key to Roman Use of the Salamis Tablet, Memorandum, 2011.
14. D. E. Smith, *History of Mathematics* (vol. II), Dover Publications, New York, 1958.
15. The Abacus: A Brief History, http://www.ee.ryerson.ca/~elf/abacus/history.html.
16. K. Menninger, *Number Words and Number Symbols: A Cultural History of Numbers*, Dover Publications, New York, 1992.
17. S. Chrisomalis, *Numerical Notation: A Comparative History*, Cambridge University Press, New York, 2010.

CHAPTER 9

Book of Calculation by Fibonacci

Leonardo Pisano Bigollo (c. 1170–c. 1250), also known as Fibonacci, was born in Pisa, Italy. He is best known as the author of *Liber Abaci* (Book of Calculation) (the first edition was published in 1202 and the second edition was published in 1228) [5]. *Liber Abaci* was one of the most important books on mathematics in the Middle Ages, introducing the Hindu numeral system (also called Hindu-Arabic numeral system or Indian-Arabic numeral system) and arithmetic methods throughout Europe [1, 2, 4, 7]. He is also well known for introducing a number sequence called *Fibonacci's numbers* (also called the *Fibonacci sequence*) to Europe [3, 8]. His extant writings about his early life are in the second paragraph of "Dedication and Prologue" in the 1228 edition of *Liber Abaci*. We quote the paragraph below from the English translation by L. E. Sigler [5]:

> As my father was a public official away from our homeland in the Bugia* customs house established for the Pisan merchants who frequently gathered there, he had me in my youth brought to him, looking to find for me a useful and comfortable future; there he wanted me to be in the study of mathematics and to be taught for some days. There from a marvelous instruction in the art of the nine Indian figures, the introduction and knowledge of the art pleased me so much above all else, and I learnt from them, whoever was

---

* Bugia is now Bejaia, a Mediterranean port in northeast Algeria, North Africa.

learnt in it, from nearby Egypt, Syria, Greece, Sicily and Province, and their various methods, to which locations of business I travelled considerably afterwards for much study, and I learnt from the assembled disputations. But this, on the whole, the algorithm and even the *Pythagorean arcs,*' I still reckoned almost an error compared to the Indian method. Therefore strictly embracing the Indian method, and attentive to the study of it, from mine own sense adding some, and some more still from the subtle Euclidean geometric art, applying the sum that I was able to perceive to this book, I worked to put it together in XV distinct chapters, showing certain proof for almost everything that I put in, so that further, this method perfected above the rest, this science is instructed to the eager, and to the Italian people above all others, who up to now are found without a minimum. If, by chance, something less or more proper or necessary I omitted, your indulgence for me is entreated, as there is no one who is without fault, and in all things is altogether circumspect.

Fibonacci was a citizen of the maritime city-state of Pisa, but he was educated in mathematics as a youth in Bugia. He continued to develop as a mathematician by traveling on business and studying in the Barbary Coast of the Western Muslim Empire, Egypt, Syria, Province, Byzantium, and other places. He learned Greek mathematics from *Euclid's Elements*, and the Hindu-Arabic numeral system (i.e., decimal place-value number system), calculation, and algebra from Arabic scientists. Although knowledge of the Hindu-Arabic numeral system began to reach Europe around the end of the 10th century, its advantage was not generally recognized in Europe even in the 12th century. Fibonacci decided to write *Liber Abaci* to bring the numeral system to the Italian people in a usable form. The effect of the book was tremendous in dissemination of the Hindu-Arabic numeral system throughout Europe [1, 2, 6, 8].

Liber Abaci consists of 15 chapters. In Chapter 1, the 10 numeral figures of the Hindu-Arabic numeral system are presented, including zero (0, 1, 2, 3, 4, 5, 6, 7, 8, 9). The zero is called zephir from the Arabic. This system is our familiar decimal place-value number system, in which the figure in the

---

' Pythagorean arcs are marks to indicate triples of place-value numbers. In the writing of numbers Fibonacci uses this number system. For example, for 1,234,567,890, each of the triples 234, 567, and 890 is covered by an arc.

first place (rightmost position) counts only for itself, but the figure in the second place to the left counts as so many tens. In the sequence of figures, the third place from the right end counts as so many hundreds, and so on. The zero denotes nothing and serves as a placeholder. Large numbers are organized by triples to facilitate our easy reading of them. For indicating these triples, Fibonacci uses "arcs," which play the same role as commas used in representing large numbers. He also showed a way of representing numbers up to 1000 by the fingers and the palm of the left hand. At the end of Chapter 1, additions and multiplications of some small numbers are given in tables. He described ways of how to carry out the additions and multiplications by using the tables, the fingers, and the palm.

In Chapter 2 an algorithm for multiplication is given. An algorithm for addition and an algorithm for subtraction of whole numbers of arbitrary size are given in Chapters 3 and 4, respectively. In Chapter 5 an algorithm for the divisions by integral numbers is given. The answer to a division is given in a form of its quotient with a fraction of its remainder (a numerator) over the divisor (a denominator). In Chapter 6 an algorithm for the multiplication with fractions is described. There, he also presents the *Euclidean algorithm* for finding the greatest common divisor of two integers. Chapter 7 continuously describes the addition, subtraction, and division of whole numbers with fractions. Fibonacci also discusses the separation of fractions into sums of unit fractions. This method goes back to the ancient Egyptian preference for unit fractions.

In Chapters 8, 9, 10, and 11, specific business and merchandise problems are given. These are problems on the buying and selling of commercial commodities, the process of bartering, certain rules for buying coins, and the alloying of money and the rules that are pertinent to alloying.

Chapter 12 is divided into nine parts. The first part is on summing series of numbers. The second is on proportions of numbers. The third is on problems of tree lengths. The fourth is on problems of finding purses. The fifth is on buying horses among company members according to given proportions. The sixth is on problems of the travelers. The seventh is on false position and rules of divination. The method of false position works by the posing of arguments that are approximations, and then the approximations are corrected to give true solutions. The eighth is on certain problems of divination. The ninth is on the doubling of squares and certain other problems. Most of these problems are equivalent to linear equations of the simple type, $Ax = B$. For example, the following is one of the tree length problems:

> There is a tree. Its $(1/4 + 1/3)$ portion lies underground, and the length of this portion is 21 palms (1 palm is about 7 to 10 cm). Find the tree length.

If we let the tree length be x palms, then the solution to the tree problem above is the solution to linear equation $(1/4 + 1/3)x = 21$. Thus, the solution to the problem is 36 palms.

Chapter 13 begins with summing arithmetic series with applications to some problems for travelers. Another one is the *found purse* problem. For example, men have purses containing some denari in each purse (*denaro* is the Pisan monetary unit, and *denari* is its plural). Conditions are given and one must find the amount of denari in each purse. Fibonacci often uses negative numbers in *Liber Abaci*. While the purses cannot contain negative numbers of denari, Fibonacci shows that they can be used to obtain solutions to the purse problem. There are also banking problems about investments with simple and compound rates of interest. The famous Fibonacci numbers (also called the Fibonacci sequence) are also included in Chapter 13. The Fibonacci sequence is described as the following story about the birth production of rabbits:

> A certain man put a pair of fertile rabbits in a place surrounded on all sides by a wall. How many pairs of rabbits can be produced from that pair in a year if it is supposed that every month each fertile pair delivers a new pair and if each new pair becomes fertile at the age of one month. [Here, we assume that the rabbits do not die.]

The resulting sequence of new pairs produced in the 1st month, the 2nd month, ..., and the 12th month is 1, 1, 2, 3, 5, 8, 13, 21, 34, 55, 89, 144. Thus, the answer to Fibonacci's question above is 144.

In Chapter 14 Fibonacci collects techniques for finding square and cube roots. In Chapter 15 he gives a review of proportion and a collection of elementary geometry problems. He uses the *Pythagorean theorem* to calculate areas and volumes of simple shapes. The techniques of algebra for quadratic equations are also presented in Chapter 15. The presentation for the techniques differs little from al-Khwarizmi's book *The Compendius Book on Calculation by Completion and Balancing* (in Arabic).

Al-Khwarizmi (c. 780–c. 850) was born in a Persian family. He was a great mathematician at the House of Wisdom in Baghdad. His *The Compendius Book on Calculation by Completion and Balancing* presents the systematic solution of linear and quadratic equations. The book was published more than 350 years earlier than *Liber Abaci*. Fibonacci studied Hindu-Arabic mathematics mainly from the work of al-Khwarizmi. In Renaissance Europe, Al-Khwarizmi was considered the original inventor of algebra, but it is now known that his mathematical work is based on older Hindu or Greek sources. The Fibonacci sequence had been also described in Hindu mathematics much earlier than Fibonacci.

Practica Geometriae is another book written by Fibonacci (1220). It contains a large collection of geometry arranged into eight chapters with theorems based on *Euclid's Elements*. Later, *Liber Quadratorum* (Book of Squares) was also written by Fibonacci (1225). In this book he proves some interesting number theoretic results. For example, he proved the following facts:

> There is no pair of integers (x, y) such that $x^2 + y^2$ and $x^2 - y^2$ are both squares.
> There is no non-trivial integral solution for $x^4 - y^4 = z^2$.

Liber Quadratorum established Fibonacci as the major contributor to number theory in the period between Diophantus and the 17th-century French mathematician Pierre de Fermat (1601–1665).

In mathematics and arts, two quantities are in *golden ratio* if the greater quantity to the smaller quantity is equal to the ratio of the sum of these two quantities to the greater quantity. In Figure 9.1, if AC:CB = AB:AC, then AC and CB are in golden ratio.

Ancient Greek mathematicians studied what we call the golden ratio because of its frequent appearance in geometry. The golden ratio has fascinated Western intellectuals of diverse interests for at least 2400 years.

In modern mathematics the first two numbers of the Fibonacci sequence are 0 and 1, although both the first two numbers of the original Fibonacci

FIGURE 9.1 Golden ratio.

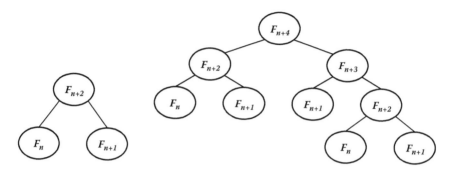

FIGURE 9.2   Examples of tree representations for Fibonacci numbers. (a) A binary tree for $F_{n+2} = F_{n+1} + F_n$ (left figure). (b) A binary tree for $F_{n+4} = 3F_{n+1} + 2F_n$ (right figure).

sequence are 1. A French-born mathematician, Albert Girard (1595–1632) [12, 13], was the first person who formulated the inductive definition of the Fibonacci sequence as:

$$F_0 = 0, F_1 = 1, F_{n+2} = F_{n+1} + F_n \qquad (n \geq 0)$$

where F_n is the nth Fibonacci number (n = 0, 1, 2, 3, ...). He stated that the ratios of two consecutive Fibonacci numbers tend to the golden ratio = $(1 + \sqrt{5})/2$. This result was published in 1634, 2 years after his death. The tree representations for equations $F_{n+2} = F_{n+1} + F_n$ and $F_{n+4} = 3F_{n+1} + 2F_n$ are shown in Figure 9.2.

Many authors attribute the discovery of the following equation to J. P. M. Binet (1786–1856) in 1843, and so call it *Binet's formula* [11]:

$$F_n = (1/\sqrt{5})(\varphi_1{}^n - \varphi_2{}^n),$$

where $\varphi_1 = (1 + \sqrt{5})/2$ and $\varphi_2 = (1 - \sqrt{5})/2$ (i.e., φ_1 and φ_2 are the roots of quadratic equation $x^2 - x - 1 = 0$). However, Leonhard Euler (1707–1783) discovered the same formula in 1765, much earlier than Binet [10]. Moreover, D. E. Knuth [9] stated in his book that the formula above can be traced back even further to a paper published in 1730 by A. de Moivre (1661–1754) [14, 15].

In modern science and technology, the Fibonacci sequence has been applied in many diverse areas. These applications include computer algorithms such as the *Fibonacci heap* data structure and interconnection networks in parallel and distributed systems called the *Fibonacci cubes*.

The Fibonacci sequence also appears in biology: branching in trees, the numbers of spirals in a pineapple, and the arrangement of a pine cone.

It is an interesting observation that the conversion factor 1.609344 for *miles to kilometers* is close to the golden ratio $(1 + \sqrt{5})/2$. Hence, the decomposition of distances in miles into a sum of Fibonacci numbers becomes nearly the kilometer sum when the Fibonacci numbers are replaced by their successors.

REFERENCES

1. Wikipedia, Fibonacci, http://en.wikipedia.org/wiki/Fibonacci.
2. Leonardo Pisano Fibonacci, http://www-history.mcs.st-and.ac.uk/Biographies/Fibonacci.html.
3. Wikipedia, Fibonacci Number, http://en.wikipedia.org/wiki/Fibonacci_number.
4. Wikipedia, *Liber Abaci*, http://en.wikipedia.org/wiki/Liber_Abaci.
5. Fibonacci (Author) and L. E. Sigler (trans.), *Fibonacci's Liber Abaci* (Leonardo Pisano's Book of Calculation), Springer-Verlag, Berlin, 2002.
6. R. E. Grimm, The Autobiography of Leonardo Pisano, *Fibonacci Quarterly*, 21, 99–104, 1976.
7. J. Hoyrup, Fibonacci—Protagonist or Witness? Who Taught Catholic Europe about Mediterranean Commercial Arithmetic? Preprint for the workshop presentation on Borders and Gates or Open Spaces? Knowledge Cultures in the Mediterranean during the 14th and 15th Centuries, Universidad de Sevilla, Sevilla, Spain, 2010.
8. G. Ifrah, *The Universal History of Computing*, John Wiley & Sons, New York, 2001.
9. D. E. Knuth, *The Art of Computer Programming* (vol. 1, 2nd ed.), Addison-Wesley, Reading, MA, 1973.
10. L. Graham, D. E. Knuth, and O. Patashnik, *Concrete Mathematics* (2nd ed.), Addison-Wesley, Reading, MA, 1994.
11. Wikipedia, Jacques Philippe Marie Binet, http://en.wikipedia.prg/wiki/Jacques_Philippe_Marie_Binet.
12. Wikipedia, Albert Girard, http://en.wikipedia.org/wiki/Albert_Girard.
13. Albert Girard, http://www-history.mcs.st-and.ac.uk/Biographies/Girard_Albert.html.
14. Wikipedia, Abraham de Moivre, http://en.wikipedia.org/wiki/Abraham_de_Moivre.
15. Abraham de Moivre, http://www-history.mcs.st-and.ac.uk/Biographies/De_Moivre.html.

CHAPTER 10

Decimal Fractions and Logarithms

10.1 APPEARANCE OF DECIMAL FRACTIONS

A decimal fraction is a decimal representation of a real number in the form of a series written as a sum:

$$r = a_0 + a_1/10 + a_2/10^2 + a_3/10^3 + \ldots \text{ if } r \geq 0$$

$$r = -a_0 - a_1/10 - a_2/10^2 - a_3/10^3 - \ldots \text{ if } r < 0,$$

where a_0 is a nonnegative integer, and a_1, a_2, ... are integers satisfying $0 \leq a_i \leq 9$ for each $i \geq 1$.

Although there is no clear evidence, it is believed that decimal fractions may have first been developed and used in China in around the first century BC, from there spreading to the Middle East [2, 3]. The textbook *Mathematical Treatise in Nine Sections* (1247 AD), by Chinese Southern Song dynasty mathematician Qin Jiushao (1202–1261), describes decimal fractions. In Europe, they probably did not appear until the 14th century. A French Jewish mathematician and astronomer, Immanuel Bonfils (c. 1300–1377), published astronomical tables and methods of decimal arithmetic including decimal fractions around the year 1350 [3].

The Persian mathematician Jamshid al-Kashi (c. 1380–1429), and the first director of the Science Institute founded in Samarkand in 1414 by

Emperor Ulugh Beg, claimed to have discovered decimal fractions during the 1420s. He computed 2π to nine sexagesimal digits, accurately converting this approximation to 17 decimal places in 1427 [4]. He gave not only the value of π more accurately than any of his predecessors, but also wrote it using Arabic numerals as follows [1]:

Integer

3   1415926535898732

The clear notation of a decimal fraction by al-Kashi seems to predate any similar notation found in Europe. In his book *The Teatise on the Chord and Sine*, al-Kashi calculated sin 1° as accurately as his calculation for π. It is also known that al-Kashi was the first person who provided a clear statement of the law of cosine (i.e., $c^2 = a^2 + b^2 - 2ab \cos \gamma$, where a, b, and c are side lengths of a triangle and γ is the angle between the sides with lengths a and b opposed to the side with length c).

The use of a dot or bar before and after integers to indicate decimal fractions became common in Europe in the 16th century. In 1585, the Flemish mathematician Simon Stevin (1548–1620) wrote *De Thiende* (*La Disme* in French, *Decimal Arithmetic* in English), a booklet first published in Flemish and later translated into French. In the booklet he clearly described the significance of the decimal fractions and demonstrated the rules for calculations involving decimal fractions as easily as if they involved only integers [1]. For example, 27.847 + 37.675 + 875.782 = 941.304 was written as 27(0)8(1)4(2)7(3) + 37(0)6(1)7(2)5(3) + 875(0)7(1)8(2)2(3) = 941(0)3(1)0(2)4(3).

By the 16th century, European mathematicians were able to produce various tables of accurate approximations for square roots, trigonometric functions, and other quantities denoted by decimal fractions. Several writers used a period to separate the decimal fraction portion from its integer portion, while others used a bar for demarkation. The improvement in notation for the decimal fractions was largely made by several scholars, including the Swiss clockmaker and mathematician Jobst Burgi (1552–1632), the German astronomer and mathematician Johannes Kepler (1571–1630), the German mathematician Johann Hartmann Beyer (1563–1625), and the Scottish mathematician John Napier (1550–1617).

10.2 LOGARITHMS

A logarithm is the inverse function of an exponential function. If $x = a^y$, then y is the logarithm of x to base a and is written $y = \log_a x$. The logarithm to base-10 is called the common logarithm (also called the decimal logarithm or Briggsian logarithm) and has many applications in science and technology. The logarithm to base e (= 2.718...) is called the natural logarithm, and it is widely used in mathematics, especially in calculus. Nowadays, the logarithm to base-2 is commonly used in computer technology and information science.

Arithmetica Integra (Integer Arithmetic), published in 1544 by the German monk and mathematician Michael Stifel (1487–1567), contained a table of integers and powers of 2. The book showed how an approximation of the multiplication of two numbers can be obtained by adding their 2' powers and looking up their sum in the table of powers of 2. Stifel's method of multiplication was based on the relation $a^m a^n = a^{m+n}$, known also to the ancient Greek mathematician Archimedes. This method can be considered an early version of multiplication by logarithms.

Logarithms are said to be first invented by John Napier in 1614, in his book *Marifici Logarithmorum Canonis Descriptio* (Description of Wonderful Rules of Logarithms). Napier dedicated at least 20 years to his work on logarithms [5, 6]. The book was subsequently translated into English in 1616. The word *logarithm*, first used by Napier, means "ratio number." His idea was to simplify multiplication involving trigonometric function values, while realizing that logarithms could also be useful for other operations. His discovery for simplifying multiplication may have been inspired by the following equation:

$$\sin A \sin B = (\cos (A{-}B){-}\cos (A + B))/2.$$

In Napier's time, there were seven-digit tables of the trigonometric functions. Example 10.1 shows how to multiply by using the tables with the operations of addition, subtraction, and division by 2.

Example 10.1

Suppose we want to calculate an approximate value of 484.8096 × 27.56374:

$$484.8096 \times 27.56374 = 0.4848096 \times 0.2756374 \times 10^5.$$

From the sine and cosine table, sin 29° is nearly equal to 0.4848096 and sin 16° is nearly equal to 0.2756374.

From sin A sin B = (cos $(A-B)$ – cos $(A + B)$)/2 and the cosine table, 484.8096×27.56374 is nearly equal to $10^5(\cos 13° - \cos 45°)/2$ and is nearly equal to $10^5(0.9743701 - 0.7071068)/2$. Then it is nearly equal to 13363.165. Since the exact value is 13363.16576, the approximate value is quite close.

Napier's logarithms (hereafter denoted by Nap.log) are not those logarithms used today. His logarithms are symbolically expressed as

$$\text{Nap.log } x = 10^7(\log_e (10^7/x)).$$

Napier demonstrated certain laws relating logarithm computations, which we can state symbolically as follows [1]:

1. If $a{:}b = c{:}d$, then Nap.log a + Nap.log d = Nap.log b + Nap.log c.

2. If $a{:}b = b{:}c$, then Nap.log a = 2 Nap.log b – Nap.log c.

3. If $a{:}b = b{:}c = c{:}d$, then 3 Nap.log b = 2 Nap.log a + Nap.log d, and 3 Nap.log c = Nap.log a + 2 Nap.log d.

Henry Briggs (1561–1630), an English mathematician and the first professor of geometry at Gresham College, London, was greatly interested in astronomy. When he read Napier's work on logarithms, he thought that logarithms would be very useful for astronomical calculations. He began studying how Napier's logarithms could be improved. Briggs traveled from London to Edinburgh to discuss logarithms with Napier in the summer of 1615 [7]. Before meeting in Edinburgh, Briggs had suggested to Napier in his letter that logarithms should be to base-10 instead of base e. Originally Nap.log 1 is not 0 by its definition. Briggs also suggested that the value should be 0. Napier had also been considering these changes. Briggs made the second journey to meet Napier in Edinburgh in 1616, and planned to make a third visit the following year, but Napier died in the spring before the planned visit. Symbolically, the later version of Napier's logarithms took the following form [1, 7]:

$$\text{Nap.log } x = 10^9 \log_{10} x.$$

Briggs's work on logarithms, *Logarithmorum Chilias Prima* (The First Thousand Logarithms), was published in 1618, and his mathematical treatise, *Arithmetica Logarithmica* (Logarithm Arithmetic), was published in 1624. In the latter book, Briggs gave the logarithms of natural numbers 1 to 20,000 and 90,000 to 100,000 that are computed in 14 decimal places.

Jost Burgi (1552–1632), a Swiss clockmaker and mathematician, invented logarithms independently of Napier and Briggs, although it is not clear when he started work on them. Some historians have suggested that Burgi may have invented logarithms earlier than Napier [1, 8], but his work was not published until 1620, when the German mathematician and astronomer Johannes Kepler asked him to do so. This date was 6 years after the publication of Napier's book.

By the middle of the 17th century, logarithms had become recognized as a useful tool for arithmetic calculation not only in Europe, but also in China, through the influence of the Jesuit missionaries [1].

In addition to inventing logarithms, Napier also invented a calculating device known as *Napier bones* in 1617 and made common the use of decimal multiplications and divisions. The device itself does not use logarithms, but rather is a convenient tool to reduce multiplication and division to a sequence of simple addition and subtraction operations. The method employed by Napier's bones was based on Arab mathematics and Fibonacci's *Liber Abaci*. The device was a set of rods, each of which was inscribed with a part of the multiplication table for the integers 1 to 9. Napier coined the word *rabdology* from the Greek words *rabdos*, meaning "rod," and *logos*, meaning "calculation," to describe this device and its technique. Napier's bones was popularly used to multiply, divide, and even find the square roots and cube roots of numbers [10], and was still used as a teaching device in British primary schools into the 20th century, see Figure 10.1.

The slide rule (also known as a slip stick in the United States) was used primarily for multiplication and division, and also for calculating square roots, cubic roots, logarithms, and trigonometric functions. The English mathematician Edmund Gunter (1581–1626) first invented it during the 1620s, shortly after Napier's publication on logarithms. The English mathematician William Oughtred (1575–1660) and others further developed the slide rule in the 17th century. Before the advent of the pocket electronic calculator, slide rules were commonly used calculation tools in mathematics, science, and engineering. By the end of the 1960s, the use of slide rules had become largely obsolete, and they were seldom manufactured [9].

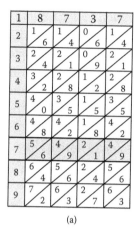

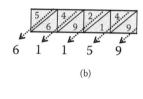

FIGURE 10.1   Napier bones and calculation of 8737 × 7. (a) Rods for 8, 7, and 3; (b) Calculation of 8737 × 7.

At present, logarithms are used in the definitions of various forms of measurement. For example, the scale of an earthquake is measured by its magnitude M, defined as

$$\log_{10} E = 4.8 + 1.5M,$$

where E is the energy (Joule) of the earthquake wave.

Another example is the measure for acidity and basicity of an aqueous solution known as pH in chemistry, defined as

$$pH = -\log_{10} a_{H^+},$$

where a_{H^+} is the hydrogen ion activity.

The decibel (denoted by dB) is also a logarithmic unit, commonly used in physics and engineering to express the ratio between two values. For example, the gain of a voltage amplifier that amplifies v_i volts to v_o volts is defined as

$$20 \log_{10}(v_o/v_i) \text{ (dB)}.$$

While the methods and tools to calculate decimal fractions and logarithms have undergone great transformations over the past four centuries, these numerical representations have become an integral part of the languages of modern science, engineering, and mathematics.

REFERENCES

1. D. E. Smith, *History of Mathematics* (vol. II), Dover Publications, New York, 1958.
2. J. Needham (ed.), *Decimal System: Science and Civilization in China, Mathematics and the Sciences of the Heavens and the Earth* (vol. III), Cambridge University Press, Cambridge, UK, 1959.
3. Wikipedia, Decimal, http://en.wikipedia.org/wiki/Decimal.
4. Wikipedia, Jamshid al-Kashi, http://en.wikipedia.org/wiki/Jamsh.
5. Wikipedia, Logarithm, http://en.wikipedia.org/wiki/Logarithm.
6. John Napier, http://www-history.mcs.st-andrews.ac.uk/Biographics/Napier.html.
7. Henry Briggs, http://www-history.mcs.st-andrews.ac.uk/Biographics/Briggs.html.
8. Jost Burgi, http://www-history.mcs.st-andrews.ac.uk/Biographics/Burgi.html.
9. Wikipedia, Slide Rule, http://en.wikipedia.org/wiki/Sliderule.
10. Wikipedia, Rabdology, http://en.wikipedia.org/wiki/Rabdology.

CHAPTER 11

Calculating Machines

The origins of mechanical computing devices are rooted in the European Renaissance. During this time, there was a resurgence of classicism and interest in humanistic subjects: thanks to the gradual spread of Indo-Arabic numeration and mathematics in the 16th century, scholars were performing increasingly complicated calculations. This movement heralded the decline of the use of Roman numerals and counter-boards, which had been predominantly used in both academia and commerce from the time of the Roman Empire. Great cities arose as a result of burgeoning trade throughout Europe, the Middle East, and Asia, and along with this trend came sufficient wealth to support the arts and sciences. Mathematicians and scientists needed more accurate computation on longer numbers in various areas than before. As their expertise grew, they searched for easier, quicker, and more reliable ways to calculate [7, 10, 16].

Tools like counting boards, Napier's bones, and the abacus, everywhere common, were not true calculating machines. While they certainly made calculation easier, they did not mechanize it. Rather, they were simply extensions of the human hand, not tools for mechanically carrying out arithmetic operations.

The key to the problem of mechanizing arithmetic operations was to find a way to reduce the human role while increasing the reliability of the results generated by the automatic movement of a machine. The search for solutions to this problem led to the invention and development of a variety of numerical calculating machines. The idea of a mechanical calculator may seem obvious now, but 400 years ago, such an idea was bold and daring, not only with regard to design, but also with regard to construction.

11.1 THE *RECHEN UHR* OR "CALCULATING CLOCK" OF WILHELM SCHICKARD

The first step in automatic calculation was taken in 1623, when the German astronomer Wilhelm Schickard (1592–1635) constructed his *Rechen Uhr*, or "calculating clock," as he called it. He was born in the small town of Herrenberk, near Tübingen, Wurttemberk (now Germany). He received a B.A. degree in 1609 and a M.A. degree in 1611 from the University of Tübingen, where he studied theology, oriental languages, mathematics, and astronomy [1–4]. Schickard first met Johannes Kepler (1571–1630), the distinguished German astronomer, around 1617. Upon meeting Schickard, Kepler was impressed with his scientific abilities and engraving skills in both wood and copperplate. Kepler wrote in his diary: "I met an excellent talent, a math-loving young man, Wilhelm, a very industrious mechanic and lover of oriental languages" [4]. He asked Schickard to assist in calculating tables, and to draw and engrave figures for his books (*Harmony of the World*: Books IV and V). Schickard's work with Kepler prompted him to design the mechanism for a calculating device.

In 1619 Schickard was appointed as professor of Hebrew at the University of Tübingen, and in 1631 he became professor of astronomy, mathematics, and geodesy. Between the years 1623 and 1624, Schickard wrote Kepler two letters, in which he sketched the design of a machine called a calculating clock. He explained in his letters that the machine could be used for calculating astronomical tables. The calculating clock could add and subtract six-digit numbers while indicating the overflow of the limit of digits by ringing a bell. We can infer from the letters that the machine could also perform multiplication and division. The first calculating clock, built by a clockmaker, was later destroyed in a fire, as described below in Schickard's letter to Kepler:

> I had placed an order with a local man for the construction of a machine for you, but when this machine had half finished, this machine, together with some other things of mine, especially several metal plates, fell victim to a fire broke out unseen during the night three days ago. I took the loss very hard, especially since there is no time to produce a replacement soon [1–4].

The design sketches of the calculating machine have been preserved in the letters to Kepler from Schickard. In 1718, a German biographer of Kepler,

Michael Gottlieb Hansch (1683–1749), published a book about Kepler, which included Schickard's two letters. These letters were rediscovered in 1935 in the Kepler archive by his biographer, Max Caspar. In 1957, Hammer, another biographer of Kepler, announced during a conference on the history of mathematics that Schickard's calculating clock predated Pascal's calculating machine, the *Pascaline*. In 1960, *Bruno v. Freytag Löringhoff*, a professor of philosophy at the University of Tübingen constructed the first replica of Schickard's machine. The Institute for Computer Science at the University of Tübingen is called the Wilhelm-Schickard Institute in his honor.

Schickard died in Tübingen, Germany, on October 24, 1635, as a result of an outbreak of the bubonic plague.

11.2 THE PASCALINE

Blaise Pascal (1623–1662), born in Clermont-Ferrand, France, was a French mathematician, physicist, inventor, and philosopher. A child prodigy, he wrote a significant treatise on the subject of projective geometry at age 16. Early on, Pascal also made important contributions to the study of pressure and vacuum in the natural and applied sciences. Later he corresponded with Pierre de Fermat (1601–1638) on probability theory. The Pascal (Pa), named after him, is commonly used as the International System Unit of pressure [5].

Pascal began to work on the design of a mechanical calculator in 1642, while assisting his father, a tax commissioner. His goal was to construct a device to reduce his father's tedious workload of calculation. Pascal built more than 20 of these calculating machines, called the Pascaline, over the subsequent 12 years. In 1649, Pascal received a royal privilege granting the right to manufacture and sell Pascalines in France. By 1654 about 20 Pascalines had been sold in Europe [6].

The Pascaline was the first calculating machine ever commercialized. Nine Pascalines have survived to the present. Seven of them are in European museums, one belongs to IBM, and one is in private hands.

The Pascaline was constructed of metal cogwheels. A dial on each cogwheel displayed a digit at the corresponding decimal position. To input a digit, the user placed a stylus in the corresponding dial, turning it until reaching a metal stop at the bottom. By repeating this process, a number was displayed in each of the boxes located across the top of the machine. The addend was similarly dialed, causing the sum of both numbers to appear in the boxes at the top. To subtract one number from another, the number to be subtracted was first converted to its nine's complement. For example, for five-digit subtraction such as 34,563–12,246, the nine's complement of 12,246 is 87,753 (where each digit of the nine's complement is the difference of 9 and the corresponding digit of the number to be subtracted). Thus, for example, 34,563–12,246 would be carried out by the addition of 34,563 + 87,753 + 1–100,000 = 22,317. (Note that for this example 100,000 would overflow.)

The Pascaline was also capable of automatic carrying. This was done by a series of cogwheels or gears, numbered from 0 to 9, and linked in such a way that when one gear completed a revolution, the next wheel was advanced by one step. In the case of subtraction, the conversion from a number to its nine's complement would be done automatically by the machine.

The functions of the Pascaline were quite limited. While multiplication and division were possible by skillful users, the machines had no special mechanisms for these purposes. Doing multiplication or division involved many steps, requiring significant effort by the user.

The production of the Pascaline ceased in 1654. By that time Pascal had turned his interests to theology and philosophy. He wrote *Letters Provincials*, a series of 18 letters on religious controversy (1656–1657), and *Pensees*, his most influential theological work, unfinished at the time of his death and published in 1669.

In 1971, Niklaus Wirth (1934–), a Swiss computer scientist, designed and published an innovative programming language that he named Pascal to honor the contributions to machine calculation made by Blaise Pascal. This programming language is suitable for writing structured programs and describing computer algorithms [18].

11.3 LEIBNIZ AND THE STEPPED RECKONER

Gottfried Wilhelm von Leibniz (1646–1716), a German mathematician, scientist, and philosopher, was born in Leipzig, Saxony, during the final stages of the Thirty Years' War (1618–1648). His excellent contributions are well known in the history of both mathematics and philosophy. Leibniz developed infinitesimal calculus independently of Isaac Newton (1642–1725), and together they are considered the founders of differential and integral calculus. Leibniz is also known as the inventor of the calculating machine called the *Stepped Reckoner* [8].

Leibniz designed the Stepped Reckoner in 1673, but constructed it later in 1694. It was the first calculating machine capable of four fundamental arithmetic operations (i.e., addition, subtraction, multiplication, and division) by purely mechanical means. Addition and subtraction were performed in a single step. Multiplication and division were performed digit by digit on the multiplier or divisor digits. The procedures for these arithmetic operations are equivalent to the familiar processes for long multiplication and long division, as taught in school nowadays. The machine can add or subtract an 8-digit number to or from a 16-digit number. It can multiply two 8-digit numbers to obtain a 16-digit result, and can divide a 16-digit number by an 8-digit divisor [9].

The Stepped Reckoner did not work well since its complicated gear-work caused major problems in construction. The tools available for machine manufacture then were not technically capable of building a reliable and robust calculating machine. Two prototypes of the Stepped Reckoner were built by Leibniz, but never sold. Today only one survives, and is displayed in the National Library of Lower Saxony in Hannover, Germany. Later, several replicas of the Stepped Reckoner were built, and exhibited in museums.

It was Leibniz who truly opened the way for the further development of mechanical calculation. Technologically, the Stepped Reckoner made many important innovations. Leibniz's contribution was critical up to the start of the computer age of the 20th century. His operating mechanism was employed in many calculating machines for the next 200 years, including the hand calculators of the 1970s.

11.4 THE JACQUARD LOOM

While the development of calculating machines discussed so far is based on the need for numerical computation, the motivation that resulted in the earliest form of a *stored program* came from a very different source: the textile industry.

One of the fundamental constructs of computational systems is the ability to represent information. In ancient times, written symbols for numeric values were developed, and then eventually, the invention of mechanical tools was inspired by these symbols. Thus, we have the arrangement of pebbles on an abacus frame, the moving scales on a slide rule, and the cogged gears on the machines of Schickard, Pascal, and Leibniz. All are examples of representational techniques that try to simplify the complex processes of arithmetic tasks. Besides numbers, however, there are other forms of information upon which computational processes can be performed. The weaving technology developed by Joseph Marie Jacquard (1752–1834) in 1801 is one example of these forms of information.

Joseph Marie Jacquard (1752–1834) was born in the famous textile city of Lyon, France. Even as a child Jacquard worked as a pull-boy on the draw looms used to produce the beautiful brocade silks worn by aristocrats and the wealthy. His father was a master weaver, and when he died in 1772, Joseph inherited his workshop and looms. By 1778, Joseph Jacquard had himself become a master weaver and silk merchant [11].

As a result of the Industrial Revolution, in the late 18th century there was a great expansion in automated industrial processes in important industries such as in textile manufacturing. Before the development of mechanical looms and weaving machines, fabric had to be tediously woven by hand. The invention of powered tools for carrying out this task meant that quantities of fabric could be mass-produced far more quickly than previously, thereby reducing its expense as well as increasing the volume of sales.

There was one technique, however, in which the new machines could not compete with skilled manual workers—that was in the production of cloth having woven patterns.

The *Jacquard loom*, the earliest programmable loom, invented by Jacquard, provided a solution to this problem. Using the Jacquard loom, very intricate patterns and pictures could be automatically woven into fabric at almost the same speed as plain fabric could be woven. The key idea behind Jacquard's loom was to control the weaving process by interfacing the action of the loom with an encoding of pattern to be woven. To do this, Jacquard represented the pattern as groups of holes *punched* into a sequence of pasteboard cards. Each card represented one weft row on the fabric. The presence or absence of a hole was detected mechanically, and the information was used to control which warp threads were raised while weft threads were interwoven. By connecting the cards in sequence on a long tape, the Jacquard loom was able to weave patterns of great complexity. A surviving example is a black-and-white silk portrait of Joseph Marie Jacquard woven under the instructions of some 24,000 cards [13].

Jacquard's invention of the punched card is important not only for the textile industry, but also for its influence on the developers of future computing machinery, including Charles Babbage (1791–1871), Ada Lovelace (1815–1852), and Herman Hollerith (1860–1929). Jacquard's invention was a hallmark in mechanization: for the first time, humans could communicate with and be understood by the machines they operated. The language of warp and weft, that is, weft threads woven over or under their warp counterparts, translated easily into the presence or absence of holes on a card. This same binary logic became our first foray into the world of programming software.

11.5 BABBAGE'S MECHANICAL COMPUTERS

By the turn of the 19th century, England was gripped by the fever of the Industrial Revolution; increasingly, people flooded into the cities to live and work in factories. New technologies of the efficiency of mass production were rapidly developing.

During the same period, both professionals and nonprofessionals alike relied heavily on a wide range of mathematical tables to perform the sorts of calculations ordinarily done on a hand calculator or a computer today. These tables were quite lengthy, and specialized tables on a specific subject sometimes filled volumes of books.

In 1790, the French mathematician and engineer Baron Gaspard de Prony (1755–1839) used the concept of the division of labor from Adam Smith's (1723–1790) *Wealth of Nations* to mass-produce tables of logarithms and trigonometric functions [15]. Trained mathematicians first worked out the formulas to produce the tables, and then broke the calculation process down into increasingly simple steps involving only simple operations. This was work that could be performed by assembly line-like workers with only a rudimentary knowledge of arithmetic. The production line involved carrying out repetitive calculations, with one person's results passed to the next group of laborers, who would then carry on the computation to the next step. The production of mathematical tables in this way became speedy; however, it came at the cost of quality: an error early on in the process was compounded at each step along the way, causing tables to be chronically incorrect.

Charles Babbage (1792–1871), an English mathematician, philosopher, and mechanical engineer, had seen Baron de Prony's tables when visiting Paris in 1819; he was greatly concerned about the errors found in mathematical tables and wanted to devise a mechanical solution to calculation that would remove the human error. Impressed by de Prony's method, he believed that de Prony's calculation work could be done by a machine.

In 1822 Babbage proposed his *difference engine*—a mechanical calculating machine, so named because it worked by adding constant differences to specified starting values. This machine would eliminate the reliance on human calculations for producing tables. His sponsors examined his drawings for the crank-driven, cogwheel difference engine and found the idea promising. They persuaded the British government to fund him

1500 pounds. Babbage predicted his difference engine would be complete within 3 years.

A working model of the difference engine proved to be an overwhelming task. Babbage spent the next 10 years modifying, enhancing, and redesigning the device. The British government advanced him a further 17,000 pounds, but ever in the search for an even better machine, Babbage abandoned work on the difference engine and undertook the design of the analytical engine in 1833. The remarkable steam-powered engine was to be an all-purpose machine: the world's first programmable computer, designed not just for solving one particular type of mathematical problem, but able to carry out any operation programmed by its operator.

Babbage was convinced that any problem, if represented numerically, could be processed mechanically. The analytical engine was composed of a *store* (now called the computer memory) and *mill* (now called the central processing unit or CPU). The computation results could either be stored or printed out. Babbage's analytical engine architecture was remarkably like modern computer memory and processors.

Just as with the design of the difference engine, Babbage continued to struggle to find an efficient and fast method of doing multiplication and division, essential in a machine as sophisticated as the analytical engine. Because these processes were so complicated, it required resetting the machinery itself each time—a laborious and time-consuming task. The final breakthrough came in 1836 when Babbage decided that instructions and data would be provided to the analytical engine by a series of punched cards, an idea he had borrowed from Joseph Jacquard and his weaving loom.

The construction of a working analytical engine, however, was not successful since the technology of the time had not yet achieved a standard sufficient for constructing the many sophisticated parts of the machine. Babbage continually adjusted his design to the frustration of his engineers. These frequent design changes were the source of constant conflict and exhausted his financial resources. The British government eventually lost faith in him since he never finished what he started—neither the difference engine nor the analytical engine. Despite many years of hard work, he realized he would never build either engine, but he continued to the end to work on their design [14].

11.6 ADA LOVELACE, THE FIRST COMPUTER PROGRAMMER

Augusta Ada King (Countess of Lovelace, 1815–1852) was the daughter of the distinguished British poet Lord Byron (George Gordon Byron,

1788–1824). In 1816, shortly after her birth, Lord Byron left his family and had no further contact with his daughter. Ada was privately schooled in mathematics and science by prominent academicians such as the mathematician and logician Augustus De Morgan (1806–1871). Ada's interest in mathematics dominated her life even after her marriage to the Earl of Lovelace and the birth of three children. Ada's understanding of mathematics was aided by her vivid imagination.

At the age of 17 Ada met Mary Fair Somerville (1780–1872), a Scottish science writer and polymath, who later introduced Ada to Charles Babbage in 1834. Learning of his intriguing design for the revolutionary analytical engine, Ada was captivated by his revolutionary ideas.

Through frequent correspondence with Babbage, Ada became one of the few to grasp the potential of his machine. Because of her familiarity with his work, Babbage suggested to Ada translating into English a French essay on the analytical engine written in 1842 by the Italian mathematician Federico Luigi Menabrea (1809–1896). In the process of doing so, Ada added copious notes, tripling the length of the original text, and making her essay, "Sketch of the Analytical Engine," a greatly enhanced description of Babbage's machine.

In her essay of 1843, Ada compared the analytical engine with the Jacquard loom. She wrote, "We may most aptly say that the Analytical Engine weaves algebraic patterns just as Jacquard's loom weaves flowers and leaves." Ada also foresaw that the analytical engine could be used to compose music, produce graphics, or be useful for other nonmathematical applications.

Ada is credited with writing a program-like algorithm for the analytical engine to calculate Bernoulli numbers. It is now widely regarded as the first computer program, thus making Ada the first computer programmer. A programming language developed by the U.S. government was named *Ada* in her honor in 1979.

After completing her essay on Babbage's analytical engine, Ada became very ill and died of cancer at the age of 36. Though her life was short, Ada's vivid imagination as the enchantress of numbers allowed her to anticipate in the 19th century many of the features of modern computing [12, 17].

11.7 HERMAN HOLLERITH AND HIS AMAZING TABULATOR

The 19th century saw great advancements in mechanizing computation thanks to the work of Charles Babbage and Ada Lovelace. While the analytical engine was never built, it opened the door to all-purpose data processing. By the end of the century, electricity had become the new darling of the Industrial Revolution as it entered an electromechanical era. Herman Hollerith (1860–1929), a mining engineer and a statistician, took mechanical computing into this next phase and into the 20th century.

The son of German immigrants, he grew up in Buffalo, New York. After receiving a Ph.D. in engineering from Columbia University, he worked first for the U.S. Census Bureau and later at MIT. While working for the Census Bureau, he learned of the dismal state of census data collection and processing. Information gathered from the 1880 census was taking years to process; with the explosive growth of the population due to successive waves of immigrants landing on American shores, prospects for the 1890 census were even more dismal. Hollerith responded to the Census Bureau's urgent need and began developing a machine that would greatly streamline data processing and enable statisticians to analyze the census data much more thoroughly.

In 1884, Hollerith applied for his first patent for a machine that used paper tape to represent information. Hollerith was well aware of the Jacquard weaving loom and its cards. His brother-in-law was in the silk weaving business, and Hollerith had often visited his mills. For this reason, he found the idea of storing information in the form of punched holes very practical. Instead of cards, however, he first planned to use paper tape since this was already commonly used by telegraph services to relay information via electric current, just as he had hoped to do. Hollerith discovered that paper tape tore easily, so he moved on to the use of the more durable cards.

In fact, Hollerith was not designing just a single machine; he was designing a whole system that would employ a standardized format for data recording, automated reading and counting, and finally, sorting. This threefold system included a card punch machine, or pantograph, named after an 18th-century device used to enlarge diagrams by mapping. To record data, the operator would indicate the paper variable on a guide-plate, and the pantograph would punch a hole in a standardized card in the appropriate place.

Punched cards were then placed on a hinged grid of blunt needles on the card reader. When the operator pulled the lever, the needles passed through the card holes into a container of mercury below, completing an electric circuit connected to dials on the machine, counting the data on the cards. Each time an electrical impulse was sent, the appropriate dial advanced incrementally.

The card reader was electrically attached to a card sorter that contained several covered bins for sorting. Depending on the kind of statistics being compiled, the sorter would automatically open the correct bin once the card had been read, and the operator would drop the card into the bin, going on to read the next card. Skilled operators could process thousands of cards a day, making the Hollerith system an immediate success. The Census Bureau was able to accurately complete the 1890 census data processing in record time, and furthermore, for the first time ever, statisticians were able to use the data to compile new and valuable statistics for the rapidly growing country, making the census an important tool for governance.

With this resounding success under his belt, Hollerith formed the Tabulating Machine Company in 1886 and was able to lease his system to other countries for their censuses. He named his machine the *Tabulator* and continued to work on refinements and new inventions. It would be fair to say that Hollerith ushered in the era of information processing at the end of the 19th century.

An engineer by trade, Hollerith was expert at neither company management nor sales. In the end, ill health convinced him to sell his business to Charles R. Flint (1850–1934), a financier also known as the Father of Trusts, who merged it with two other companies, forming the Computer Tabulating Recording (CTR) Company in 1911. A few years later, Flint hired Thomas J. Watson (1874–1956), a former executive of NCR Corporation, to become the general manager. With Watson's keen business and sales acumen, CRT grew steadily. Hollerith continued to serve as a consulting engineer for the company until 1921. In 1924, the company was renamed International Business Machines (IBM). Herman Hollerith died in 1929 after a lifetime of remarkable engineering achievements, and having achieved the successful application of Jacquard's marvelous cards.

IBM's punched cards were eventually standardized at 7 and 3/8 inches by 3 and 1/4 inches. In 1929, IBM began using rectangular holes instead of round ones. Each card contains 12 rows of 80 columns, and each column is typically used to represent a single piece of data. The 80-column card dominated the punched card market from around 1950. Although punched cards are rarely used now, they had a great influence on the computer industry during the 1950s and 1960s.

REFERENCES

1. Wilhelm Schickard, School of Mathematics and Statistics, University of St. Andrews, http://www-history.mcs.st-andrews.ac.uk/Biographies/schickard. html.
2. Wikipedia, Wilhelm Schickard, http://en.wikipedia.org/wiki/Wilhelm_ Schickard.
3. *Wilhelm Schickards Tübinger Rechenmaschine von 1623* (ed.), Bearbeitet von Friedrich Seck., Tübingen, 2002.
4. The Calculating Clock of Wilhelm Schickard, History of Computing, http:// history-computer.com/MechanicalCalculators/Pioneers/Schickard.html.
5. Wikipedia, Blaise Pascal, http://en.wikipedia.org/wiki/Blaise_Pascal.
6. Wikipedia, Pascal's Calculator, http://en.wikipedia.org/wiki/Pascal's_calculator.
7. M. R. Williams, *History of Computing Technology*, IEEE Computer Society, Los Alamitos, CA, 1997.
8. Wikipedia, Gottfried Leibniz, http://en.wikipedia.org/wiki/Gottfried_Leibniz.
9. Wikipedia, Stepped Reckoner, http://en.wikipedia.org/wiki/Stepped_Reckoner.
10. G. Ifrah, *The Universal History of Computing*, John Wiley & Sons, New York, 2001.
11. Wikipedia, Joseph Marie Jacquard, http://en.wikipedia.org/wiki/Joseph_ Marie_Jacquard.
12. J. Palfreman and D. Swade, *The Dream Machine*, BBC Books, London, 1991.
13. J. Essinger, *Jacquard's Web: How a Hand Loom Led to the Birth of the Information Age*, Oxford University Press, New York, 2004.
14. D. Swade, *The Difference Engine: Charles Babbage and the Quest to Build the First Computer*, Penguin Books, New York, 2002.
15. B. Collier and J. MacLachlan, *Charles Babbage and Engines of Perfection*, Oxford University Press, New York, 1998.
16. H. Blohm, S. Beer, and D. Suzuki, *Pebbles to Computers, the Thread*, Oxford University Press, Toronto, 1986.
17. B. A. Toole, *Ada, the Enchantress of Numbers, Prophet of the Computer Age*, Strawberry Press, Mill Valley, CA, 1998.
18. N. Wirth, The Programming Language Pascal, *Acta Informatica*, 1, 25–63, 1971.

CHAPTER 12

Solutions to Algebraic Equations

Being able to solve algebraic equations is at the heart of many modern computer applications. Whether one is simply converting a Celsius temperature to Fahrenheit, processing simultaneous linear equations, or working with complex, higher-degree equations, finding the solutions to algebraic equations is a basic feature of many computer applications. Algebraic equations come in many forms and are normally identified by the maximum degree of their terms. Finding the general solution for algebraic equations has been an endeavor that dates back to the early Chinese, Indian, and Arab mathematicians. As any high school mathematics student knows, some classes of algebraic equations have general solutions; for example, the quadratic equation $ax^2 + bx + c = 0$ has the famous formula:

$$x = \frac{-b \pm \sqrt{b^2 - 4ac}}{2a}.$$

As we will see, very interesting people, some with tragically short lives, made the many contributions that allow us today to know which algebraic equations have general solutions and which do not. Having these general solutions to certain algebraic equations makes it much easier to write computer software that solves a given algebraic problem in a finite amount of time.

12.1 LINEAR EQUATIONS

A linear equation is a sum of terms, where each term is either a constant or a product of a single variable and a constant called a coefficient. For example, $2x + 4 = 0$ is a linear equation in a single variable x, and $y = 2x - 3$ is a linear equation in two variables x and y. The latter one is also called a linear function (i.e., y is a linear function of x). On the other hand, for example, both $x^2 + 1 = 0$ and $2x + y^7 - 6 = 0$ are not linear equations, nor is $xy + 4 = 0$.

Evidence indicates that many early civilizations had learned how to solve simple linear equations. Clay tablets, from Mesopotamia, dated 1800 to 1600 BC, show methods for solving linear and quadratic equations. Egyptian mathematical texts from approximately this same time period show how to solve first-order linear equations. It is worth noting that early Chinese mathematicians had also mastered linear equations. By the year 200 BC, Chinese were able to solve systems of linear equations with two unknowns.

Today we frequently rely on algorithms based on the Gauss–Jordan method to solve systems of linear equations, and this brings us to Johann Carl Friedrich Gauss (1777–1855). Around 1810, Gauss became interested in computing the orbit of a newly discovered asteroid in our solar system, Pallas. Using observations, he was able to describe the behavior of Pallas using six linear equations and six unknowns. To solve this system of equations, Gauss invented his method of Gaussian elimination. His method involves performing row operations on the linear equations, arranged as a matrix, to form an upper-triangular matrix. When in this form, the last equation has one unknown and can then be solved. This solution is then inserted in the second-to-last equation, to form another one unknown equation, and so on.

Example 12.1

Consider the following system of linear equations:

$$\begin{cases} x + 2y - z = 1 \\ -2x - y + 2z = 1 \\ 3x - 4y - z = 1 \end{cases}.$$

We show how to find the solution to this system of linear equations by the Gaussian elimination method. The augmented matrix of these linear equations is

$$\begin{pmatrix} 1 & 2 & -1 & 1 \\ -2 & -1 & 2 & 1 \\ 3 & -4 & -1 & 1 \end{pmatrix}.$$

Adding twice the first row to the second row, and subtracting three times the first row from the third row, we have

$$\begin{pmatrix} 1 & 2 & -1 & 1 \\ 0 & 3 & 0 & 3 \\ 0 & -10 & 2 & -2 \end{pmatrix}.$$

We multiply the second row by 1/3 and add 10 times the second row to the third row. Then we multiply the third row by 1/2. These processes are

$$\begin{pmatrix} 1 & 2 & -1 & 1 \\ 0 & 1 & 0 & 1 \\ 0 & -10 & 2 & -2 \end{pmatrix} \rightarrow \begin{pmatrix} 1 & 2 & -1 & 1 \\ 0 & 1 & 0 & 1 \\ 0 & 0 & 2 & 8 \end{pmatrix}$$

$$\rightarrow \begin{pmatrix} 1 & 2 & -1 & 1 \\ 0 & 1 & 0 & 1 \\ 0 & 0 & 1 & 4 \end{pmatrix}.$$

Adding the third row to the first row and subtracting twice the second row the from the first row, we have

$$\begin{pmatrix} 1 & 2 & 0 & 5 \\ 0 & 1 & 0 & 1 \\ 0 & 0 & 1 & 4 \end{pmatrix} \rightarrow \begin{pmatrix} 1 & 0 & 0 & 3 \\ 0 & 1 & 0 & 1 \\ 0 & 0 & 1 & 4 \end{pmatrix}.$$

From the last matrix, $x = 3$, $y = 1$, $z = 4$.

12.2 QUADRATIC EQUATIONS

In mathematics, a quadratic equation is a polynomial equation of the second degree. The general form is $ax^2 + bx + c = 0$, where x represents a variable or an unknown, and a, b, and c are constants with $a \neq 0$. (If $a = 0$, the equation is a linear equation.)

Our look at quadratic equations starts in India. The Indian mathematician Brahmagupta (598–668 AD) is thought to have been born in Bhinimal, a city in northwest India, and lived under the patronage of King Vyaghramukha. He is often referred to as Bhillamalacarya, loosely translated to "the teacher from Bhillamala." Interestingly, his approach to computation included the then-new notions of zero and negative numbers. It is believed that Brahmagupta's famous book, *Brahmasphutasiddhanta* (Corrected Treatise of Brahma), was translated in the 12th century to Arabic, and this book's numerical system formed the early basis for the Arabic numerals, which eventually made their way to Europe and modern mathematics. One of Brahmagupta's main contributions was in Chapter 18 of *Brahmasphutasiddhanta*, the solution of the general linear equation.

Although the terminology he used was different, Brahmagupta had correctly denoted the quadratic equation, which we now know as:

$$x = \frac{-b \pm \sqrt{b^2 - 4ac}}{2a}.$$

Another interesting personality is the Persian mathematician and scholar Muḥammad ibn Mūsā al-Khwārizmī (780–850 AD). He was a scholar at the Baghdad House of Wisdom, studying science and mathematics. Translations of his work, during the 12th century, brought the modern number system, using decimal positions, to the Western world. al-Khwārizmī was considered the Father of Algebra in Renaissance Europe; however, we now know that his approach to numbers originated from much older sources, such as Brahmagupta in India.

It is thought that the famous computer science term *algorithm* is derived from the Latinization of his name, Algoritmi. In addition, the term *algebra* is credited to al-Khwārizmī, and his book *Al-Kitāb al-mukhtaṣar fī hīsāb al-ğabr wa'l-muqābala* (The Compendious Book on Calculation by Completion and Balancing) provides a process for solving polynomials of up to degree 2. As today's modern mathematical notations were not known, the book explains the quadratic solution processing using ordinary text.

12.3 CUBIC EQUATIONS

A cubic equation is a polynomial equation of the third degree. The general form is $ax^3 + bx^2 + cx + d = 0$, where x represents a variable or an unknown, and a, b, c, and d are constants with $a \neq 0$. (If $a = 0$, the equation is a quadratic equation at most.)

With how to solve quadratic equations understood around 900 AD, we now move on to cubic equations, 600 years later, and Renaissance Italy. Here we find an interesting story of discovery, involving four very capable mathematicians: Scipione del Ferro (1465–1526), who first discovered a general solution to cubic equations; Nicolo Fontana Tartaglia (1500–1557), who a few years later independently discovered a similar general solution; Girolamo Cardano (1501–1576), who first published a general solution; and Lodovico Ferrari (1522–1565), who is described as a student of Cardano.

This would be a simple story if Scipione del Ferro had published his approach to finding the solution to general cubic equations. However, in del Ferro's Italy it was common for mathematicians to publicly challenge one another to solve each other's problems, with the loser losing prestige and possibly his academic position. So, it was also common for mathematicians to hold their discoveries secret, and use them to defeat other mathematicians, if challenged. For whatever reason, del Ferro never published his approach. After his death, his approach was only discovered when Girolamo Cardano and Lodovico Ferrari received del Ferro's notebooks, containing his approach.

This would all be well and good if it were not for the fact that Nicolo Fontana Tartaglia had also discovered a similar approach to solving cubic equations—and had shown it to Cardano, with the understanding that Cardano would not publish his approach. When Cardano found del Ferro's preexisting approach, Cardano decided that the commitment he had made to Tartaglia was no longer valid, and he published del Ferro's approach in his 1545 book *Ars Magna*. This infuriated Tartaglia, and what followed was a decade-long fight, with Tartaglia publicly insulting Cardano. Cardano's student, Lodovico Ferrari, took up the defense of his teacher. Ferrari leads us to our next topic (see below), solving quartic and quintic equations.

The approach Cardano published, actually found by del Ferro and Tartaglia, is rather complex for our purposes but involves doing a reduction to get a *depressed cubic*. This depressed cubic is of the form $t^3 + pt + q = 0$, where p and q are equal to new equations expressed totally in terms of the constants a, b, c, and d from the original cubic equation.

12.4 QUARTIC AND QUINTIC EQUATIONS

A quartic equation is a polynomial equation of the fourth degree. The general form is $ax^4 + bx^3 + cx^2 + dx + e = 0$, where x represents a variable or an unknown, and a, b, c, d, and e are constants with $a \neq 0$. (If $a = 0$, the equation is a cubic equation at most.) A quintic equation is a polynomial equation of the fifth degree. The general form is $ax^5 + bx^4 + cx^3 + dx^2 + ex + f = 0$. As we will see, some very intelligent and also tragic young mathematicians showed that quintic equations do not have a general *algebraic solution* (i.e., solutions in radicals). An algebraic solution means a solution of an algebraic equation in terms of the coefficients relying on addition, subtraction, multiplication, division, and the extraction of roots (i.e., radicals). For example, $x^5 - x + 1 = 0$ is a quintic equation such that its roots cannot be expressed in terms of radicals.

Let us continue with Lodovico Ferrari, the student of Cardano. Ferrari responded to Tartaglia's attacks on Cardano, and in Milan in 1548 there was a public debate between Ferrari and Tartaglia, concerning the solution of algebraic equations. It seems the younger Ferrari had a much deeper understanding of the mathematics, and the older Tartaglia soon refused to engage the younger Ferrari and left under the cover of darkness, ceding the debate to Ferrari. In turns out that Ferrari was a very capable mathematician in his own right, and among Ferrari's accomplishments is finding the general solution to quartic equations [3].

We now turn our attention to the general solution of quintic equations, these being equations of degree 5. Whereas with the previous equation types, that is, linear, quadratic, cubic, and quartic, we had success in having general algebraic solutions, this area is less satisfying and tragic. It is less satisfying because we learn there is no general algebraic solution to quintic equations utilizing rational numbers and radicals, and tragic because of the deaths that befell two of the young mathematicians that led the way to these quantic findings.

But first, we describe the contributions of Joseph Louis Lagrange (1736–1813). Lagrange was a mathematician, astronomer, and academic who lived at various times in Prussia and France. His doctoral adviser was none other than Leonhard Euler, and among his doctoral students were Joseph Fourier and Simeon Poisson. Lagrange's *Theorie des fonctions analytiques* set some of the foundational work for Galois groups, and we will now meet Evariste Galois (1811–1832) and Niels Henrik Abel (1802–1829).

Niels Henrik Abel was a famous Norwegian mathematician, born in Nedstandrad, Norway, the second of seven children, to Søren Georg Abel and Anne Marie Simonen. Although born into a relatively prosperous and large family, by the time Abel went to the Royal Frederick University, he was nearly destitute due to the untimely death of his father.

At age 16, Abel produced a proof of the binomial theorem for all numbers, extending Euler's earlier work. At age 19, he proved that there is no general algebraic solution for the roots of quantic equations, this being his *impossibility theorem*. His proof relied on group theory, which he invented independently of Galois. After his death, it was discovered that Abel had also produced an impressive work on elliptic functions.

Abel was able to travel, due to the patronage of others, through France and Germany, where he met with Legendre and others. Through a series of missteps and bad luck, Abel failed to get his works fully recognized. However, in Berlin he met and befriended August Crelle (1780–1855), an amateur mathematician who published a journal on pure and applied mathematics. Crelle would go on to publish several of Abel's works. Abel returned to Norway, continuing his wretched existence, having failed to find a position and support, living in extreme poverty.

Abel's mathematical career ended tragically with his death at age 27, due to his poverty and associated tuberculosis. Abel had been striving for years to obtain a position as a professor, which he badly needed to lift himself out of poverty. Ironically, a letter from Crelle arrived 2 days after his death, stating that he had been offered a professorship in Berlin. In his short life Abel achieved so much of the highest order that he was one of the leading mathematicians of the day. Charles Hermite (1822–1901) could say without exaggeration, "Abel has left mathematicians enough to keep them busy for five hundred years." Abel, having been asked how he had accomplished so much as his age, replied, "By studying the masters, not the pupils" [4].

Abel's impossibility theorem states that there is no general algebraic solution, meaning a solution in radicals, to every polynomial equation of degree 5 or higher. Although they cannot be expressed with radicals, they can be numerically computed using Newton's root finding method.

Évariste Galois (1811–1832) was a French mathematician who was born in Bourg-la-Reine. While still in his teens, Galois was able to determine the exact conditions that permit some polynomial equations to be solved by radicals. In particular, he showed that there is no general solution for quintic polynomial equations and was the first to use the term *group* to refer to a group of permutations.

Galois entered the Lycée Louis-le-Grand School at age 12 and performed well for the first 2 years but became bored with his studies. At age 14, he became seriously interested in mathematics, and it is said he read Marie Legendre's *Éléments de Géométrie* like it was a novel, that is, in one reading. By age 15, he was reading the original papers of Lagrange.

However, his teachers found his academic work uninspiring. His inability to show his abilities resulted in his inability to gain acceptance to École Polytechnique, the most famous French mathematics institution of the day. He settled for acceptance to École Normale, a lesser institution. About this time, he published his first paper on continued fractions.

Galois lived at a time of political turmoil in France, and he got caught up in the politics of the day. This resulted in his being expelled from his school, École Normale. Leaving school, he eventually joined an artillery unit of the National Guard and became friends with the unit's officers. After the arrest of the unit's officers, it was disbanded and he returned to his study of mathematics. However, at a ball where the officers were present, Galois got manipulated into a pistol duel over a love affair.

Early on a morning in May 1832, participating in the pistol duel, Galois was shot in the abdomen and died the following day, at age 20. The night before the duel, suspecting that he might be killed, he wrote his scientific "testament" in the form of a letter to a friend. In it, he referred to some of his unpublished discoveries. These discoveries included the theory of groups and Galois's theory of equations, which established the limits on generalized solutions to algebraic equations. Galois was able to produce criteria that showed which algebraic equations are solvable and which are not [1, 2].

Galois's most important contribution to mathematics is his development of what is now called Galois theory. His theory originated in his search for why quantic (fifth- or higher-degree) polynomials lack general solutions using only algebraic operations and radicals. All the polynomial types up to quartic have solutions of this type. At the heart of Galois's theory is the notion of considering the composition of the permutations of roots. These roots yield groups of polynomials of a lesser degree.

Although the detailed specifics of theory are beyond our scope, we can look at a simple example with a quadratic equation. Let us note that this quadratic equation,

$$x^2 - 4x + 1 = 0$$

has these two roots:

$$A = 2 + \sqrt{3}$$

$$B = 2 - \sqrt{3}.$$

We can further note that these can be combined into two valid equations:

$$A+B = 4$$

$$AB = 1.$$

After looking at the possible permutations, using only algebraic operators and radicals (no irrational numbers, for example), we see that there are only two equations; all others are isomorphic. Knowing that there are two is sufficient to show that these are the roots of this quadratic equation. Évariste Galois's contributions made it possible to know with certainty which polynomials are solved by radicals and which are not.

REFERENCES

1. L. Infeld, *Whom the Gods Love: The Story of Evariste Galois*, McGraw-Hill Book, New York, 1948.
2. J.-P. Tignol, *Galois' Theory of Algebraic Equations*, World Scientific Publishing, Singapore, 2001.
3. D. E. Smith, *History of Mathematics* (vol. II), Dover Publishing, New York, 1958.
4. Abel, Niels Henrik (1802–1829), from Eric Weisstein's World of Scientific Biography, Scienceworld.wolfram.com, http://en.wikipedia.org/wiki/Niels_Henrik_Abel, which links to http://scienceworld.wolfram.com/biography/Abel.html (accessed July 12, 2011).

CHAPTER 13

Real and Complex Numbers

13.1 REAL NUMBERS

Roughly speaking, a *real number* is a value that represents a quantity on the number line, and any point on the number line represents a real number. That is, the set of real numbers can be thought of as the set of all points on an infinitely long number line. Once we decide that a point on the line represents *zero*, any point to the right from the zero point and any point to the left from the zero point represent a positive real number and a negative real number, respectively. The set of integers is properly included in the set of rational numbers, and the set of rational numbers is properly included in the set of real numbers.

Any two numbers can be ordered on the number line. If point A locates to the left of point B, then A is smaller than B (i.e., $A < B$). If $A < B$ and $B < C$, then $A < C$. Therefore, ordering among real numbers is transitive. Any set with this property is called a *totally ordered set*. The set of integers, the set of rational numbers, and the set of real numbers are all totally ordered. Points representing consecutive integers on the number line locate discretely, in equally spaced intervals. They look like stepping stones in a garden. There is no integer between any pair of consecutive integers. On the other hand, between any pair of distinct rational numbers $a < b$, there exists a rational number r such that $a < r < b$. The *density* of rational numbers signifies this property, as does the set of real numbers. The set of integers is, by definition, not dense.

Any number between two real numbers is also a real number, but between any pair of distinct real numbers there exists a real number such that it is not a rational number. A real number that is not rational is called an irrational number. This property of real numbers is said to be *continuous*. In other words, the set of real numbers is continuous, but the set of rational numbers is not continuous on the number line.

The concept of irrational numbers appeared among Hindu mathematicians around 600 BC. Around 500 BC Greek mathematicians, probably including Pythagoras, realized that irrational numbers must exist. For example, geometrically $\sqrt{2}$ is the length of the diagonal across a square with sides of 1 unit length. They noticed that such a length cannot be represented by a rational number [1].

In the Middle Ages, Arabic mathematicians treated irrational numbers as algebraic objects. In the 17th century French philosopher and mathematician Rene Descartes (1596–1659) introduced the term *real number* to describe the roots of a polynomial equation. A real number that is a root of a nonzero polynomial with rational coefficients is called an *algebraic* (real) number. All rational numbers are algebraic, but not vice versa. There exist irrational numbers that are algebraic numbers. For example, $\sqrt{2}$ and $\sqrt{3}$ are algebraic irrational numbers. The set of algebraic real numbers is a proper subset of the set of real numbers. A real number that is not an algebraic number is called a *transcendental* (real) number.

In the 18th and 19th centuries there was much work on irrational and transcendental numbers. The most prominent examples of transcendental numbers are π (the ratio of the circumference of a circle to its diameter) and e (the Napier constant). Almost all real numbers are transcendental numbers, but it is extremely difficult to show that a given number is transcendental. In 1794 the French mathematician Adrien-Marie Legendre (1752–1833) gave a complete proof demonstrating that π cannot be rational, nor can it be the square root of a rational number. The French mathematician Joseph Liouville (1809–1882) showed in 1840 that e cannot be a root of any quadratic equation with integer coefficients, and then established the existence of transcendental numbers. The German mathematician Georg Cantor (1845–1918) gave a much simpler proof of the existence of transcendental numbers in 1873 from the cardinalities of the set of algebraic equations and the set of real numbers [2] (see also Chapter 14).

Before the 1860s, descriptions of real numbers were not rigorous. In the latter half of the 18th century, there was a movement to establish a rigorous and logically sound foundation for mathematics [4]. A number of mathematicians realized that a rigorous definition of real numbers was needed as well. They are Karl Weierstrass (1815–1897), Julius Dedekind

(1831–1916), and Georg Cantor, along with others. Dedekind introduced the notion known as the *Dedekind cut* in his book in 1872. A Dedekind cut c is a partition on the number line into two nonempty parts such that every point of the first part (set A) locates to the left of any point of the second part (set B), where $A = \{a \mid a < c\}$ and $B = \{b \mid c \le b\}$. If B contains the smallest rational number, the cut c defines a rational number. Otherwise, it defines an irrational number. On the other hand, both Weierstrass and Cantor defined an irrational number by the limit of an infinite sequence of rational numbers. The rigorous construction of an irrational number in this way was first presented during lectures in the 1860s by Weierstrass. However, he never published his construction of an irrational number in a complete form [4]. We give here Cantor's definition of irrational numbers, which was given in 1872:

> Let $a_1, a_2, \ldots, a_n, \ldots$ be an infinitive sequence of rational numbers. For an arbitrarily given small positive number ε, if there exists a sufficiently large integer N such that the difference between any a_n and a_m satisfying $n, m > N$, is less than ε (i.e., $|a_n - a_m| < \varepsilon$), then this infinite sequence has its limit. If this limit is not rational, then the limit is irrational. In this way an irrational number can be defined.

In 1873 the French mathematician Charles Hermite (1822–1901) first proved that e is transcendental, and in 1882 the German mathematician Ferdinand von Lindemann (1852–1939) showed that π is transcendental. Lindemann's proof was simplified by Weierstrass in 1885, and was further simplified by David Hilbert (1862–1943) in 1893. Lindemann was the supervisor for the doctoral thesis of Hilbert at the University of Könisberg.

"Squaring the circle" is a problem proposed by the ancient Greek mathematicians. It had been an open problem whether it was possible to construct a square with the same area as a given circle using only a finite number of steps with a compass and straightedge. This problem was proven to be impossible in 1882, when Lindemann showed that π is transcendental. The hierarchy of number sets is depicted in Figure 13.1.

It seems difficult for school boys and girls to understand why π is an irrational number. The first chapter of the well-known science fiction novel *Contact* by Carl Sagan (1934–1996) [6] is "Transcendental Numbers." In it, a school math teacher explains to some seventh graders, "π is about 22/7, about 3.1416. But actually, if you want to be exact, it was a decimal that went on and on forever without repeating the pattern of numbers." Ellie,

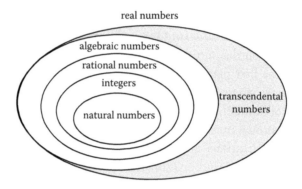

FIGURE 13.1   Hierarchy and containments of sets of numbers.

one of the girls in the class, asked the teacher, "How could anybody know that the decimals go on forever? How can you count forever?" Ellie went to the library later, and then found that the problem is very difficult and that this fact was discovered only about 250 years ago [6]. More discussion on π is found in Chapter 19.

13.2 COMPLEX NUMBERS

Some algebraic equations have solutions that are not real numbers. For example, no real number can be the solution for $x^2 + 1 = 0$. The imaginary unit i is defined to be one of the solutions of $x^2 + 1 = 0$. Another solution of this equation is $-i$. An *imaginary* number is a number that can be written as the multiplication of a real number and the imaginary unit i. A *complex* number is a number expressed in the form $a + ib$, where a and b are real numbers. The real part and the imaginary part of $a + ib$ are a and b, respectively. Real numbers can be thought of as complex numbers with their imaginary parts being zero.

As far as we know, the square root of a negative quantity appeared for the first time in the *Stereometria* of Heron of Alexandria (c. 10–70 AD) [5, 10]. Heron of Alexandria was an ancient Greek mathematician who gave the correct formula for calculating the height of a frustum of a pyramid with a square base. The height is calculated by the square root of $c^2 - 2((a - b)/2)^2$, where a and b are the edge lengths of the bottom and top squares, respectively, and c is the slant edge length. Heron used $a = 28$, $b = 4$, and $c = 15$, giving the square root of -63. Unfortunately, he seemed to misunderstand this mysterious quantity [5, 10]. Neither Greek mathematicians nor Arab mathematicians seemed to pay much attention to the subject of the square root of a negative number.

Italian mathematician Luca Pacioli (1445–1517) stated in his book *Summa de Arithmetica, Geometria, Proportioni, et Proportionalita* (1494) that the quadratic equation $x^2 + c = bx$ cannot be solved unless b^2 is not less than $4c$. From this fact, he probably recognized the impossibility of finding a real number that is equal to the square root of a negative number [10]. Italian mathematician Girolamo Cardano (1501–1576) was the first to use the square root of a negative number in an actual computation. He showed that $5 \pm \sqrt{-15}$ are the solutions to the problem of dividing 10 into two parts whose product is 40. German mathematician Gottfried Wilhelm Leibniz (1642–1716) studied imaginary numbers. However, he did not grasp the idea about geometrical representation of complex numbers.

The use of imaginary numbers was not widely accepted until the 18th century. Caspar Wessel (1745–1818) was the first person to describe the geometric interpretation of complex numbers as points on the complex plane (1799). Since his idea was published in Danish, it was not noticed by major European mathematicians [9]. The same result was later independently rediscovered by Jean-Robert Argand (1768–1822) in 1806 and Carl Johann Friedrich Gauss (1777–1855) in 1831. The complex plane has an x-axis (the real axis) and a y-axis (the imaginary axis) orthogonal to the x-axis. The real part and the imaginary part of a complex number are represented by displacements along the x-axis and the y-axis, respectively. One of the most prominent results by Wessel was the vector representation of complex numbers. He claimed that a geometrical representation of complex numbers, with length and direction, was useful for the addition of complex numbers. The vector representation of complex number $a + ib$ is the directed arrow from point $(0, 0)$ to point (a, b) in the complex plane. His idea for adding complex numbers is the same as the vector addition technique used today, see Figure 13.2.

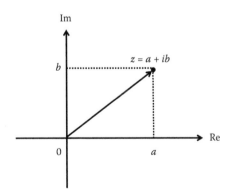

FIGURE 13.2 An example of complex plane and complex number $a + ib$.

French mathematician Jean Le Rond d'Alembert (1717–1783) first showed that every polynomial has a complex root, but his proof was not rigorous. Gauss provided a fully rigorous proof of this result. This is now called the *fundamental theorem of algebra*, or the *d'Alembert–Gauss theorem*. The set of complex numbers is closed under the arithmetic operations (addition, subtraction, multiplication, and division). This result was also shown by d'Alembert.

Example 13.1

The addition, subtraction, multiplication, and division of two complex numbers $z_1 = a + ib$ and $z_2 = c + di$ ($c \neq 0$ or $d \neq 0$) are defined in the following table:

Operation	Definition
Addition	$z_1 + z_2 = (a+c) + i(b+d)$
Subtraction	$z_1 - z_2 = (a-c) + i(b-d)$
Multiplication	$z_1 \times z_2 = (ac - bd) + i(ad + bc)$
Division	$z_1 \div z_2 = \dfrac{a+ib}{c+id} = \dfrac{(ac+bd) + i(bc - ad)}{c^2 + d^2}$

13.3 COMPLEX-VALUED FUNCTIONS

A complex-valued function (also called a complex function) is a function that may assign a complex number to each member of its domain. The domain may contain complex numbers. In the 18th century some European mathematicians turned their attention to complex-valued functions. They extended the domains of real-valued functions to the set of complex numbers and allowed their function values (range) to be complex numbers. The English mathematician Roger Cotes (1682–1716) discovered in 1714 the following formula using a complex logarithm:

$$\log_e (\cos x + i \sin x) = ix.$$

The Swiss mathematician Leonhard Euler (1707–1783) obtained the following formula using a complex exponential function instead of a complex logarithm around 1740:

$$e^{ix} = \cos x + i \sin x.$$

This is *Euler's formula*, named after Leonhard Euler. This formula establishes an important relationship between trigonometric functions and complex exponential functions. American physicist Richard Feynman (1918–1988) called Euler's formula Euler's jewel and said that it is the most remarkable formula in mathematics (1977). A proof of Euler's formula is based on the following power series (Taylor series) expansions of e^x, cos x, and sin x:

$$e^x = 1 + \frac{x}{1!} + \frac{x^2}{2!} + \frac{x^3}{3!} + \cdots$$

$$\cos x = 1 - \frac{x^2}{2!} + \frac{x^4}{4!} - \frac{x^6}{6!} + \cdots$$

$$\sin x = \frac{x}{1!} - \frac{x^3}{3!} + \frac{x^5}{5!} - \frac{x^7}{7!} + \cdots .$$

Substituting ix for x in the expansion of the exponential function above and using the expansions of the trigonometric functions above, we obtain Euler's formula.

One of the most interesting complex-valued functions is $f(x) = e^{ix}$. The locus of e^{ix} with the real number domain is the circle of radius 1 at center 0 in the complex plane. On the other hand, the locus of real function $x^2 + y^2 = 1$ is the circle of radius 1 at center (0, 0) in the Euclid plane. Noting the similarity of these loci, American mathematician and electrical engineer Charles Proteus Steinmetz (1865–1923) proposed a complex number representation for the calculation of alternating currents in an electrical circuit around 1893 [5, 11]. As a result of Steinmetz's breakthrough work, electrical engineers realized the great advantage in the use of complex quantities rather than trigonometric functions for calculating all problems of alternating circuits.

Complex-valued function theory has continued to develop from the 18th century until today, and is now a very prosperous area in mathematics, and there is an enormous amount of literature related to it. Although very mathematical in nature, there are many applications for complex-valued function theory in the field of physics and engineering.

REFERENCES

1. Wikipedia, Real Number, http://en.wikipedia.org/wiki/Real.
2. Wikipedia, Transcendental Number, http://en.wikipedia.org/wiki/Transcendental.
3. Wikipedia, Complex Number, http://en.wikipedia.org/wiki/Complex.

4. J. C. Tweddle, Wierstrass's Construction of the Irrational Number, *Math Semester*, 58, 47–58, 2011.

5. P. J. Nahin, *An Imaginary Tale: The Story of the Square Root of Minus One*, Princeton University Press, Princeton, NJ, 1998.

6. Carl Sagan, *Contact*, Simon and Schuster, New York, 1985.

7. Wikipedia, Imaginary Number, http://en.wikipedia.org/wiki/Imaginary.

8. Wikipedia, Complex Number, http://en.wikipedia.org/wiki/Complex.

9. Wikipedia, Caspar Wessel, http://en.wikipedia.org/wiki/Caspar.Wessel.

10. D. E. Smith, *History of Mathematics* (vol. II), Dover Publications, New York, 1958.

11. C. P. Steinmetz, Complex Quantities and Their Use in Electrical Engineering, presented at Proceedings of the International Congress, Chicago, 1893.

CHAPTER 14

Cardinality

One cannot begin to discuss the notion of *cardinality* without mentioning sets. Philosophers and mathematicians have always used sets, e.g., alphabet = {a, b, c, ..., z}, N = the set of natural numbers, Q = the set of rational numbers, R = the set of real numbers, etc. With sets, a natural question arose: how many elements does a given set have? While this appears to be a relatively easy question to answer, that is not quite so. Consider a set of elements where it is difficult, if not impossible, to determine if a specific element is or is not a member. How does one count the elements of such a set? Another apparently easy question is: do sets A and B have the same number of elements? Rephrased, the question is: do sets A and B have the same cardinality? If both sets have a finite number of elements, the answer is straightforward: just count the number of elements in each set and check if it is the same number. But what if the sets are not finite, as is the case with N and R. While it is obvious that N is a proper subset of R, and, in fact, there are infinitely many elements in R that are not in N, does that mean that R has more elements? How does one define a set's cardinality if the set is infinite?

The great German mathematician Georg Cantor (1845–1918) formalized the concept of sets and their cardinalities at the end of 19th century. Before Cantor, it was the opinion of most mathematicians that cardinalities of all infinite sets are just that, infinite. Cantor argued that if two sets are to have the same cardinality, then there must exist (an invertible) one-to-one function (bijection) between their respective elements.

When two sets A and B are said to be of the same cardinality, this is usually denoted by $A = {}_c B$, $cd(A) = cd(B)$, or $|A| = |B|$. There was no problem with this when the sets were finite. Problems started when Cantor applied this to infinite sets.

It was Cantor's 1874 article "Über eine Eigenschaft des Inbegriffes aller reellen algebraischen Zahlen" (On a Property of the Collection of All Real Algebraic Numbers) that was the first to prove that there was more than one kind of infinity [1]. Previously, all infinite sets/collections had been implicitly assumed to be equinumerous. Due to the novelty of Cantor's idea about cardinality, most mathematicians of the time were very critical of his work, some rejecting it outright.

Cantor defined *countable* sets as those that are finite, or can have their elements *enumerated*, i.e., put into a one-to-one correspondence with the elements of the set N. Today, these sets are also called *denumerable*. The cardinality of the set of natural numbers, which is the smallest infinite set, was assigned the symbol $\aleph_0$ (read: aleph-zero) by Cantor. He showed that there existed an indexing scheme showing a one-to-one correspondence between the set of rational numbers and N. This was possible because the rational numbers (see Chapter 3) can be represented as ratios of two integers. What was not intuitive is that the table of such ratios, say p/q, had infinitely many rows and columns. This, however, implied that the product $\aleph_0 \times \aleph_0$ was itself equal to $\aleph_0$. This was too much for some of the mathematicians to handle.

One can easily show that the set of odd integers has the same cardinality as the set of even integers; here the simple (invertible) bijection between the two sets is $f(n) = 2n$. An alternative (but which is also equivalent) argument proving that two sets A and B have the same cardinality is to show that there exist two injective functions: between A and B as well as B and A. Since it is easy to show that these injections exist, the odd, even, and their union, the set of natural numbers N, all have the same cardinality of $\aleph_0$. It follows that $\aleph_0 + \aleph_0 = \aleph_0$. Other counterintuitive arithmetical operations, e.g., a polynomial in $\aleph_0$ is equal to $\aleph_0$, can be shown.

Another example showing the less than intuitive consequences when operating with infinite sets is highlighted in the following example: let A be an infinite countable set. (1) Is it possible to take away from A an infinite subset and still have A remain nonempty? The answer is, of course, yes (natural numbers – odd numbers = even numbers). But now, we have a follow-up question: (2) can the process in (1) be repeated arbitrarily many times? The answer is, again, affirmative. Take the two-dimensional table corresponding to the set of rational numbers (it is of size $\aleph_0 \times \aleph_0$), but that quantity is equal to $\aleph_0$. Taking away the first column, which has infinitely many elements, will still leave us with $\aleph_0$ elements; taking away the second column, …, etc.

However, Cantor's proof that the set of real numbers contains more than $\aleph_0$ elements was even more counterintuitive, and at first was not even accepted by many of the mathematicians. The proof goes as follows:

Cantor's Proof

Let us assume that N and R have the same cardinality. Then by definition there must exist a one-to-one onto function f (i.e., a bijection) that maps each element from N to a unique element in R. Now, if such a function existed, then certainly one would also exist that mapped N to $S = \{$irrational numbers between 0 and 1$\}$. Such irrational numbers in S, e.g., $\pi/4.0$, would have the following representation in decimal: $0.d_1d_2d_3\ldots$, where the sequence of digits d_i would be both infinite and nonrepeating (if that were not the case, then the number, by definition, must be rational).

Since f is a bijection, then every element in S must have a corresponding element in N that mapped to it; in other words, it must have an index. In that case, the elements (numbers) in S can be listed as $s_1, s_2, s_3, \ldots$, and while we cannot tell exactly which number from S got to be the first one, or the second, etc., we do know that every element in S must get an index from N. We will now construct a number z that will obviously be in the set S, but which will differ from each and every number that has already received an index. The number z will have the form $0.e_1e_2e_3 \ldots$, where the ith digit past the decimal place of z will differ from the ith digit of s_i. We can do this easily by setting $e_i = s_{ii} + 1$, where s_{ii} is the ith digit of s_i. The addition is modulo 10, so $9 + 1 = 0$ (modulo 10).

Now, z is in S, but it will not receive an index. If it did get one, say k, then we have $z = s_k$, but the kth digit of z is, by construction, different from the kth digit of s_k, which is a contradiction; hence, our assumption that a bijection f between N and S existed is incorrect, thus the sets N and S must have different cardinalities. Since S has at least as many elements as N, our conclusion is that the cardinality of S must be more than $\aleph_0$. ■

Cantor's proof technique can also be used to show that for any set A, the cardinality of its *powerset*, which is the set of all subsets of A, usually denoted by P(A), is always greater than that of A itself. Today this is known as Cantor's theorem. This means that there cannot exist a universal set, i.e., a set of "everything," since its powerset will probably have more elements.

This was discovered by Cantor in 1899 in his *eponymous paradox*: what is the cardinal number of the set of all sets? Clearly it must be the greatest possible cardinal. Yet for any set A, the cardinal number of the powerset of A is strictly larger than the cardinal number of A, so there cannot be a greatest cardinality.

The cardinality of the powerset of N was assigned the symbol $\aleph_1$, and each subsequent powerset would receive an $\aleph$ with an appropriate index. Since the concept of constructing powersets has no bounds, this means that there are an infinite number of infinities. Therefore, instead of counting sheep, one might fall asleep quicker if one were to count different infinities.

The method used in the proof above is called Cantor's diagonal argument. Cantor also proved that the cardinalities of the set of complex numbers and the set of real numbers are the same. The work of Cantor did not stop here. He spent the rest of his life trying to prove what is known as the *continuum hypothesis*, which, stated very informally, reads: there is no set whose cardinality is strictly between $\aleph_0$ and $\aleph_1$. However, he failed in his quest. In 1900, the famous German mathematician David Hilbert included the continuum problem in his famous list of open problems in mathematics. In fact, it was listed as problem number 1. A partial answer was provided in 1939 by Kurt Gödel (1906–1978), who proved that the continuum hypothesis is consistent with the axioms of set theory; in other words, one cannot prove it to be incorrect. It wasn't until 1963 that an American mathematician, Paul Cohen (1934–2007), showed that the continuum hypothesis is independent of the axiom of set theory, meaning that it cannot be proven to be true or false just by using the axioms.

An example of a set whose cardinality exceeds even $\aleph_1$ is the set of all real-valued functions, i.e., mappings from R to R. The cardinality of this set is denoted by $\aleph_2$. While the cardinal numbers, which are associated with counting, can be produced by, e.g., taking powersets of previously constructed ones and whose cardinalities are denoted by $\aleph$'s with higher and higher index numbers, there exist *ordinal numbers*, usually denoted by ω, that would not be reached using this approach.

Ordinals were introduced by Cantor in 1883 to accommodate infinite sequences and to classify sets with certain kinds of structures on them (e.g., order). Here, the arithmetic is even more confusing, as, for example, ordinals are *not commutative*, so $1 + \omega$ is ω rather than $\omega + 1$ [5]. When two sets U and V are said to be of the same ordinal type, this is usually denoted by $U =_o V$.

In his paper, Cantor described "ordinal types" just a few pages after defining cardinal numbers [4]:

> Every ordered set U has a finite "ordinal type," ... which we will denote by U'. By this we understand the general concept which results from U if we only abstract from the nature of the elements u, and retain the order or precedence among them. Thus, the ordinal type U' is itself an ordered set whose elements are units which have the same order of precedence amongst one another as the corresponding elements of U, from which they are derived by abstraction.... A simple consideration shows that two ordered sets have the same ordinal type if, and only if, they are similar, so that of the two formulas $U =_o V$, $U' = V'$, one is always a consequence of the other.

The naïve definition of a set as a collection of objects has led to numerous paradoxes. The best known among these is *Russell's paradox*, which can be informally stated as: let S be the set of all sets that are not members of themselves [2, 3]. Now, a given set, say S, is either a member of itself or not. *If not* a member of itself, then by definition it must contain itself, i.e., S as a member, and if it *does* contain itself, then by definition, it shouldn't (because it already contains itself as a member).

Cantor's and Russell's paradoxes showed that the naive, or nonaxiomatic, set theory often lead to contradictions. This led to several axiomatizations, but even these were incomplete, leaving problems such as the continuum hypothesis unanswered.

One of the most important consequences of Cantor's theorem is that there must exist problems for which no algorithms (computer programs) can be written (see Chapter 27). The supporting argument is cardinality based and it goes as follows:

> Problems, e.g., determining if a number is prime or a perfect square, etc., are really functions from N to N. Computer programs actually take numbers (binary strings) as input, and output their answers (again, binary strings). The number of functions from

N to N, or equivalently, the number of problems, is the same as the number of subsets of N, which is the powerset of N. Its cardinality was shown to be $\aleph_1$. However, the number of possible programs that have been, or can ever be, written cannot exceed $\aleph_0$. So, using a straightforward cardinality argument, we have shown that there must exist problems/functions for which a computer program cannot be written.

Many practical problems in computer science, ranging from the determination of running time of an algorithm to proving the optimality or even the correctness of an algorithm, are often resolved using cardinality-based analyses.

REFERENCES

1. Wikipedia, George Cantor, http://en.wikipedia.org/wiki/George_Cantor.
2. Thomas, Jech, *Set Theory* (3rd ed.), Springer, New York, 2002.
3. Wikipedia, Russell's Paradox, http://en.wikipedia.org/wiki/Russell%27s_paradox.
4. Yiannia Moschovakis, *Notes on Set Theory*, Springer, Yew York, 1994.
5. Wikipedia, Ordinal Numbers, http://en.wikipedia.org/wiki/Ordinal_numbers.

CHAPTER 15

Boolean Algebras and Applications

George Boole (1815–1864) was an English mathematician and a founder of Boolean logic. He was born in Lincoln, Lincolnshire, England, in 1815. Boole's father, John Boole, was a dilettante in the field of science. He loved participating in discussions and lectures on science and technology, and became the curator of the library of the Lincoln Mechanics Institute. George Boole could access excellent books, and essentially taught himself foreign languages, classics, Christian theology, and mathematics [1, 2].

After 3 years working as a schoolteacher, Boole opened his own school at age 19 in Lincoln. He worked as a schoolmaster for 15 years until 1849, when he became the first professor of mathematics at Queen's University in Cork, Ireland (now University College Cork).

Boole introduced the algebra of logic in his book *Mathematical Analysis of Logic* (1847) [5, 8]. It was designed to provide an alternative, as some modification of ordinary algebra, to the traditional approach of Aristotelian logic. He developed general methods to greatly extend Aristotelian logic. Boole proposed that logical propositions should be expressed as algebraic equations. The algebraic manipulation of the symbols in the equations provides a method of logical deduction. Boole replaced the operation of addition by the word *OR* and multiplication by the word *AND*. The symbols in the equations can stand for collections of objects or statements in logic [2]. Further statements of his work on the algebra of logic were given in his book *An Investigation of the Laws of Thought* (1854) [9].

TABLE 15.1   Algebraic Laws of Set Operations

Idempotent laws	(1a)	$A \cup A = A$	(1b)	$A \cap A = A$
Associative laws	(2a)	$(A \cup B) \cup C = A \cup (B \cup C)$	(2b)	$(A \cap B) \cap C = A \cap (B \cap C)$
Commutative laws	(3a)	$A \cup B = B \cup A$	(3b)	$A \cap B = B \cap A$
Distributive laws	(4a)	$A \cup (B \cap C) = (A \cup B) \cap (A \cup C)$		
	(4b)	$A \cap (B \cup C) = (A \cap B) \cup (A \cap C)$		
Identity laws	(5a)	$A \cup \phi = A$	(5b)	$A \cap U = A$
	(6a)	$A \cup U = U$	(6b)	$A \cap \phi = \phi$
Involution laws	(7)	$(A^C)^C = A$		
Complement laws	(8a)	$A \cup A^C = U$	(8b)	$A \cap A^C = \phi$
	(9a)	$U^C = \phi$	(9b)	$\phi^C = U$

Algebraic laws of set operations are listed in Table 15.1, where $\cup$ is the symbol for the set union, $\cap$ is the symbol for the set intersection, A^C means the complement of set A, U means the universal set, and ϕ means the empty set. Both sets and logical propositions satisfy similar laws. In the propositional logic, $\vee$ and $\wedge$ are used as the symbols for the logical *OR* operation and logical *AND* operation, respectively. The symbols $\vee$ and $\wedge$ are often called the *disjunction* and the *conjunction*, respectively. The negation of proposition A is denoted by $\neg A$. Binary operations + and · are often used to mean $\vee$ and $\wedge$, respectively. Operations $\cup$ and $\cap$ in Table 15.1 correspond to $\vee$ and $\wedge$, respectively, and the complement operation corresponds to the negation operation (i.e., A^C corresponds to $\neg A$). Then the algebraic system of logical propositions satisfies the laws given in Table 15.1. Therefore, the algebra of sets and the algebra of propositions are similar algebraic systems. The laws listed in Table 15.1 can also be used to define a mathematical structure called *Boolean algebra*, which is named after George Boole. The Boolean algebra can be considered to be a variant of ordinary elementary algebra differing in its values, operations, and algebraic laws [6].

Boole's algebra predated the modern development in abstract algebra and mathematical logic. It is the algebra of the truth value and the false value, equivalently the algebra of just two subsets, U and ϕ (e.g., for the set of subsets of $\{a\}$, $U = \{a\}$ and ϕ). This Boolean algebra is denoted by B_2. In the late 19th century, Boole's work was generalized and refined by William Stanley Jevons (1835–1884), Augustus De Morgan (1806–1871), Charles Sanders Peirce (1839–1914), and William Ernest Johnson (1858–1931). Boole's algebra reached the modern concept of an abstract mathematical structure. It can be explained in the algebra of sets. In an abstract setting, Boole's algebra B_2 is isomorphic to one of the algebra of sets (i.e., Boolean algebras). In fact, Marshall Harvey Stone (1903–1989) proved in 1936 that every Boolean algebra is isomorphic to the algebra of sets. Therefore, there exists the Boolean algebra of n elements if n is a power of 2 (i.e., $n = 2^k$ for some nonnegative integer k). The Boolean algebra with n elements is denoted by B_n. More formally, a Boolean algebra is defined by the following definition.

Definition 15.1

A Boolean algebra B_n is a set of elements $a_1, a_2, \cdots, a_n$ in B_n with three types of operations, $\wedge(AND)$, $\vee(OR)$, and $\neg(NOT)$, satisfying the properties given in Table 15.1, where any of A, B, and C can take any element in B_n. ▪

Example 15.1

Consider the set of subsets of $\{a, b, c\}$. These subsets are ϕ, $\{a\}$, $\{b\}$, $\{c\}$, $\{a, b\}$, $\{a, c\}$, $\{b, c\}$, and $\{a, b, c\}$. The algebra of these subsets with the three types of set operations (i.e., set union, set intersection, and complement) is a Boolean algebra. This Boolean algebra is denoted by B_8.

De Morgan (1806–1871) was a British mathematician and logician. He studied mathematics at Trinity College, Cambridge University. He was appointed to first professor of mathematics at London University (now University College, London University). In 1842, Boole started a correspondence with De Morgan, and they became close friends. De Morgan formulated De Morgan's laws and introduced the term *mathematical induction* [3].

In propositional logic and Boolean algebra, De Morgan's laws are two related transformation rules that make it possible for one to express conjunctions exclusively in terms of disjunction, and disjunctions exclusively in terms of conjunction in logical proofs. That is, De Morgan's laws can be given as follows:

$$\neg(A \vee B) = (\neg A) \wedge (\neg B) \text{ and } \neg(A \wedge B) = (\neg A) \vee (\neg B)$$

where $\neg$ is the negation operator (*NOT*), $\wedge$ is the conjunction operator (*AND*), $\vee$ is the disjunction operator (*OR*), and $=$ means logically equivalent.

Example 15.2

The negation of a statement "He is American or British" is logically equivalent to a statement "He is not American and not British" by De Morgan's laws.

Although De Morgan's laws are named after Augustus De Morgan, a similar observation was made by Aristotle (384–322 BC) and was known to ancient Greek and medieval logicians (e.g., William of Ockham (1288–1348), an English philosopher). De Morgan's laws can be easily proved and may seem to be trivial. Nonetheless, these laws are helpful in making valid influences in proofs, deductive arguments, and equivalent transformations of logical formulae.

John Venn (1834–1923) was a British logician and philosopher. He was born in Hull, Yorkshire, England. He began his education at Sir Roger Cholmley's School (now known as Highgate School in London). He enrolled in Gonville and Caius College in Cambridge in 1853, and graduated from Cambridge University in 1857. Venn's main area of interest was logic, and he published three textbooks on the subject. He extended Boole's mathematical logic. Venn introduced the frequency interpretation of probability in *The Logic of Chance*, which was published in 1866. He introduced Venn diagrams in *Symbolic Logic*, which was published in 1881. Venn diagrams are used to show possible logical relations among a finite collection of sets. They are used to teach elementary set theory, as well as to illustrate simple set relations in probability, statistics, linguistics, and computer science. The diagram shown in Figure 15.1 is an example of a Venn diagram. The stained glass window of a Venn diagram is displayed in the dining hall of Gonville and Caius College, Cambridge University [7].

Expressions built up from the $\wedge$, $\vee$, and $\neg$ operations, Boolean constants, and any number of variables with any proper usage brackets are called

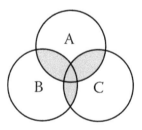

FIGURE 15.1   A Venn diagram for $(A{\wedge}B){\vee}(A{\wedge}C){\vee}(B{\wedge}C)$.

Boolean polynomials or Boolean expressions. A Boolean function or switching function $f(x_1,\cdots,x_n)$ is defined to be a mapping from $\{0, 1\}^n$ to $\{0, 1\}$. A Boolean function can be expressed by a Boolean expression. For each Boolean function, there are many Boolean expressions to express it. For example, Boolean expressions $\neg(x \wedge y)$ and $(\neg x) \vee (\neg y)$ express the same Boolean function as shown in De Morgan's laws. Boolean function values 1 and 0 are called *true* and *false*, respectively. The function value of a Boolean function $f(x_1,\cdots,x_n)$ depends upon the values of its variables $x_1,\cdots,x_n$. A simple concise way to show this relationship is through a truth table.

Example 15.3

Let us consider a game of tossing three coins by a player. If all the coins are heads or all the coins are tails, then the player gains score 1, and otherwise the player gains score 0. The score of a player can be expressed by the following Boolean function:

$$f(x_1,x_2,x_3) = (x_1 \wedge x_2 \wedge x_3) \vee ((\neg x_1) \wedge (\neg x_2) \wedge (\neg x_3)),$$

where each of variables x_1, x_2, and x_3 takes 0 (tail) or 1 (head), and the function value is the score of the player. Table 15.2 is the truth table of the Boolean function.

Any n-variables Boolean function can be realized by a combinatorial circuit with n input terminals and a single output terminal, in which OR gates, AND gates, and NOT gates are used as logical elements. The OR gate, AND gate, and NOT gate are each a Boolean function. Any of these gates can be

TABLE 15.2 Truth Table of a Boolean Function

x_1	x_2	x_3	$f(x_1,x_2,x_3)$
0	0	0	1
0	0	1	0
0	1	0	0
0	1	1	0
1	0	0	0
1	0	1	0
1	1	0	0
1	1	1	1

TABLE 15.3   Truth Tables of the *NOR* Gate and *NAND* Gate

x	y	$NOR(x,y)$	$NAND(x,y)$
0	0	1	1
0	1	0	1
1	0	0	1
1	1	0	0

expressed by using only *NOR* gates, and also can be expressed by using only *NAND* gates. The *NOR* gate is a Boolean function $f(x_1,x_2)=\neg(x_1 \vee x_2)$, and the *NOT* gate is a Boolean function $f(x_1,x_2)=\neg(x_1 \wedge x_2)$. The truth tables of the *NOR* gate and the *NAND* gate are given in Table 15.3.

Example 15.4

Let the *NOR* of x and y, and the *NAND* of x and y be expressed by *NOR(x,y)* and *NAND(x,y)*, respectively. Then the *OR* gate, *AND* gate, and *NOT* gate can each be expressed by using only *NOR* gates, as well as by using only *NAND* gates, as shown below:

$$x \vee y = NOR(NOR(x,y), NOR(x,y)) = NAND(NAND(x,x), NAND(y,y))$$

$$x \wedge y = NOR(NOR(x,x), NOR(y,y)) = NAND(NAND(x,y), NAND(x,y))$$

$$\neg x = NOR(x,x) = NAND(x,x).$$

Claude Elwood Shannon (1916–2001) was an American mathematician and electrical engineer known as the Father of Information Theory. He graduated from the University of Michigan in 1936 with two bachelor's degrees, one in electrical engineering and one in mathematics. Then he graduated from MIT with a M.S. in electrical engineering in 1937 and a Ph.D. in mathematics in 1940. He later became a research fellow at the Institute for Advanced Study at Princeton University and worked for Bell Laboratories. Eventually, Shannon returned to MIT as a professor [4].

While a graduate student at MIT, Shannon was advised by Vannevar Bush (1890–1974) and employed to maintain Bush's differential analyzer, an analog computing device. To operate this device, Shannon had to manually configure its gears, a laborious task. Bush suggested to Shannon that he might write his master's thesis on the subject of the logical operation of the differential analyzer. It occurred to Shannon that the machine could be improved if it were operated by electric circuitry rather than mechanical gears [4].

While pondering how to redesign the differential analyzer, Shannon recalled having taken a course in symbolic logic, where he was introduced to Boolean algebra. It now occurred to him that Boolean principles lent themselves most conveniently to the design of switching circuits, and that this logic could be further used to perform digital calculations. After developing this idea, he wrote his master's thesis, "A Symbolic Analysis of Relay and Switching Circuits," in 1937, and its journal paper appeared in 1938 [10]. These papers contain series-parallel two-terminal circuits, multiterminal and non-series-parallel networks, synthesis of networks, and illustrative examples.

Shannon's master thesis and its published version [10] are now recognized as the foundation of modern switching theory. The contribution of these papers was remarkable for the growth of both the computer industry and the telephone industry.

During the early stage of the computer industry, research and development of switching theory and its applications to the design of digital circuits were very active. A lot of research papers and many textbooks on switching theory with applications were published in the 1950s and 1960s, e.g., see [11–16].

The Karnaugh map is a graphical technique for representing Boolean functions for a few variables and obtaining simplified Boolean expressions. This graphical technique is due to Maurice Karnaugh (1924–) [19]. From 1952 to 1966, Karnaugh worked at Bell Telephone Laboratory where he developed pulse code modulation (PCM) encoding and magnetic logic circuits. He later worked with IBM, studying multistage interconnection networks [17].

If a Boolean function is expressed as a disjunction of one or more product terms, where each product term is the conjunction of several variables or complements (negations) of variables in which the same variable does not appear more than once, the expression is called a *disjunctive normal form*. Similarly, a dual of the disjunctive normal form is called a conjunctive normal form. The dual of a Boolean expression is the Boolean expression obtained by changing the disjunctive operator with the conjunctive operator and the conjunctive operator with the disjunctive operator. The normal form minimization problem was first studied by Willard van Orman Quine (1908–2000) [19]. The problem was also studied by Edward J. McCluskey [18]. The minimization method is called the Quine-McCluskey algorithm.

Efficient implementation of Boolean functions is a fundamental problem in the design of combinatorial logic circuits. Modern design automation tools for very-large-scale integration (VLSI) circuits rely on efficient representation of Boolean functions.

REFERENCES

1. Stanford Encyclopedia of Philosophy, George Boole, http://plato.stanford. edu/entries/boole/.
2. Wikipedia, George Boole, http://en.wikipedia.org/wiki/George_Boole.
3. Wikipedia, Augustus De Morgan, http://en.wikipedia.org/wiki/Augustus_ De_Morgan.
4. Wikipedia, Claude Shannon, http://en.wikipedia.org/wiki/Claude_Shannon.
5. Stanford Encyclopedia of Philosophy, The Algebra of Logic Tradition, http:// plato.stanford.edu/entries/algebra-logic-tradition/.
6. Wikipedia, Boolean Algebra, http://en.wikipedia.org/wiki/Boolean_algebra.
7. New World Encyclopedia, Venn, John, http://www.newworldencyclopedia. org/entry/John_Venn.
8. G. Boole, *The Mathematical Analysis of Logic, Being an Essay Towards a Calculus of Deductive Reasoning*, Macmillan, Cambridge, UK, 1847.
9. G. Boole, *An Investigation of the Laws of Thought on Which Are Founded the Mathematical Theories of Logic and Probabilities*, Macmillan, London, 1854.
10. C. E. Shannon, A Symbolic Analysis of Relay and Switching Circuits, *Transactions of the American Institute of Electrical Engineers*, 57, 713–723, 1938.
11. M. Phister Jr., *Logical Design of Digital Computers*, John Wiley & Sons, New York, 1958.
12. M. A. Harrison, *Introduction to Switching and Automata Theory*, McGraw-Hill, New York, 1965.
13. R. E. Miller, *Switching Theory: Combinatorial Circuits* (vol. 1), John Wiley & Sons, New York, 1965.
14. S. H. Caldwell, *Switching Circuits and Logical Design*, John Wiley & Sons, New York, 1958.
15. R. A. Higonnet and R. A. Grea, *Logical Design of Electrical Circuits*, McGraw-Hill, New York, 1958.
16. W. S. Humphrey Jr., *Switching Circuits with Computer Applications*, McGraw-Hill, New York, 1958.
17. M. Karnaugh, The Map Method of Synthesis of Combinatorial Logic Circuits, *Communications and Electronics*, 9, 593–599, 1953.
18. E. J. McCluskey Jr., Minimization of Boolean Function, *Bell System Technical Journal*, 35, 1417–1444, 1956.
19. W. V. Quine, The Problem of Simplifying Truth Functions, *American Mathematical Monthly*, LXII, 627–631, 1952.

CHAPTER **16**

Computability
and Its Limitations

Until the mid 1930s the notion of computability had not yet been mathematically well established, which is natural since the first electronic computers were not constructed until the *Colossus*, *ENIAC*, and *EDVAC* were built in the 1940s. It is quite interesting that the computational limitations of these (as well as all future) computers had already been proven mathematically in 1936.

16.1 GÖDEL'S INCOMPLETENESS THEOREM

At the turn of the 19th century, the German mathematician David Hilbert (1862–1943) set out to find an algorithm for determining the truth or falsehood of any mathematical proposition. Hilbert believed that for any formal mathematical theory there must exist a procedure (i.e., an algorithm) by which its provability could be decided. This became known as Hilbert's *Entscheidungsproblem* (decision problem). However, in 1931, Kurt Gödel (1906–1978), a young mathematician at the University of Vienna, published his now famous paper "Über formal unentscheidbare Sätze der Principia Mathematica und verwandter Systeme I" (On Formally Undecidable Propositions of Principia Mathematica and Related Systems I), proving that no such procedure can exist [3].

Gödel showed that the axiomatic method itself unfortunately possesses certain inherent limitations. The incomplete theorem tells us that if a (computable) system is powerful enough to describe the arithmetic of natural numbers, then it cannot be fully axiomatized. For such a system, the consistency of the axioms cannot be proved within the system. Gödel's incompleteness theorem was astonishing and attacked a central problem in the foundations of logic and mathematics. Naturally, it became important to formally define what was meant by the term *algorithm* or the phrase "effective procedure for computing a function."

Gödel's incompleteness theorem is closely related to several results regarding undecidable problems in recursion theory, the foundation of theoretical computer science. Initially, Gödel's incompleteness and subsequent related theorems left the door slightly open; i.e., there was still hope that it might be possible to produce a general procedure that would indicate whether a given statement is undecidable or not, thus allowing us to bypass the undecidable statements in the first place. The negative answer to Hilbert's Entscheidungsproblem indicated implicitly that it is impossible to decide algorithmically whether statements in arithmetic are true or false.

16.2 TOTAL FUNCTIONS

A function is a mapping between two sets, usually called the domain and the range, such that each element in the domain is mapped to at most one element in the range. If every element of the domain is mapped to exactly one element of the range, then such a function is called a total function, that is, everywhere *defined*. For example, the real function $f(x) = 2x$ is a total function. On the other hand, the function $f(x) = 1/x$ is not total, since it is undefined at $x = 0$.

An interesting question arises whether a one-to-one correspondence exists, i.e., a bijection, between N, the set of nonnegative integers, and the set of total functions from N to N. The answer to this question is no, even if we restrict our functions to those functions from N to $\{0, 1\}$. This can be verified using Cantor's famous diagonalization argument (see also Chapter 14).

Assume that the set of all functions from N to $\{0, 1\}$ has the same cardinality as N. It follows that we can enumerate these functions, i.e., assign a unique index to each function. Let f_i be the ith function in our enumeration. Now, let us construct a function g as follows: $g(i) = f_i(i) + 1$ (mod 2). Clearly g cannot be the first function f_1, since $g(1) \neq f_1(1)$. In fact, by construction, for any finite index q, our function g cannot be the same as f_q. However, since g is a total function from N to $\{0, 1\}$, then, by assumption, it should have been one of the enumerated functions, i.e., received an index. That is a contradiction, so the set of total functions from N to $\{0, 1\}$ cannot be enumerated, and therefore its cardinality exceeds that of N.

This conclusion has very profound implications. One cannot give a distinct name to each function, since the set of finite length sequences on an alphabet (i.e., some finite set of symbols) has the same cardinality as that of the set of integers. It follows that there are too many functions to give a distinct name to each one. As a consequence, functions must exist that are not computable by any program or algorithm. Observe that the set of programs or algorithms is enumerable, although it is infinite, because we can easily assign a unique name/index to each (finite length) program by simply ordering them lexicographically.

16.3 TURING MACHINES

Alan M. Turing (1912–1954) was born on June 23, 1912, in London. He graduated with distinction from Cambridge University in 1934 and was elected to a fellowship at King's College. In 1935, he attended a course on formal logic given by M. H. A. Newman (1897–1984). That course ended with a full treatment of the proof of Gödel's incompleteness theorem, which in a sense refuted the possibility of finding the algorithm that Hilbert implicitly proposed [7].

Turing was attracted by both Hilbert's challenge and Gödel's work. He noticed that Gödel did not explicitly define the meaning of mechanical operations and the mechanical process of solving a problem. He felt that a formal definition of the intuitive notion of solving a problem via such means was needed. Turing considered that such a formal definition itself was fundamental and necessary in order to understand and address these problems.

In his 1936 paper entitled "On Computable Numbers with an Application to the Entscheidungsproblem" [1], Turing introduced and gave a construction for an abstract model for a computation machine, now known as the *Turing machine* [3–6]. Turing displayed great originality in his formalization of the intuitive notion of an effective procedure in terms of discrete operations on his abstract computing machine. He demonstrated the existence of functions that were not computable and, in this way, resolved Hilbert's Entscheidungsproblem in the negative. He also defined a *computable function* as a function that can be computed by a Turing machine (which does not necessarily have to halt for every given input).

Formally, a Turing machine consists of a finite state control (i.e., a finite set of instructions) and an arbitrarily long tape. The arbitrarily long tape

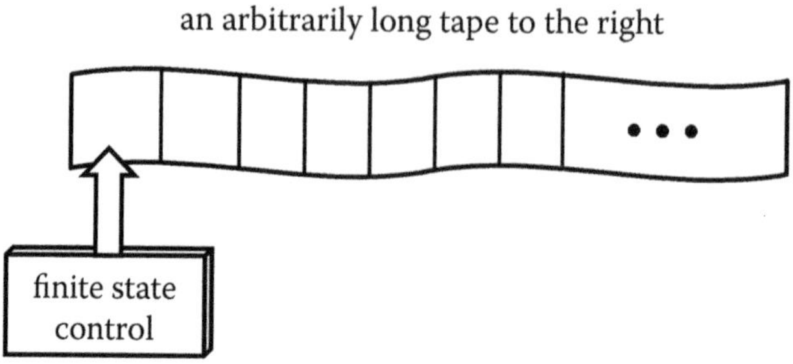

an arbitrarily long tape to the right

FIGURE 16.1   An image of a Turing machine.

means that the machine never runs out of the tape or reaches the right end of the tape. The tape, which functions as the machine's memory, is divided into squares (cells). Each square may be read or inscribed with a single symbol from a designated finite alphabet, or it may be blank. The tape has a leftmost square, but it is unlimited to the right. The control unit can shift the read-write head (to the left or right) during each step of the computation. At any single time step, the read-write head is able to examine just the one tape square over which it is positioned as shown in Figure 16.1.

The control unit of a Turing machine can assume any one of a fixed number of states that, together with the symbol read from the tape, determines how the machine will behave at that step. The actions available to a Turing machine are quite limited. It may either halt, thereby terminating its computation, or carry out a basic move. Each move consists of writing a symbol in the currently scanned tape square, shifting the read-write head one square to the right or left, and causing the control unit to enter into some new state (i.e., go to the next specified instruction).

The computation of a Turing machine proceeds as follows. The machine is initially supplied with a tape on which a finite number of squares are inscribed with symbols and the rest are left blank. The initial content of the tape is called the *input*. The read-write head is positioned over the leftmost input symbol—usually the leftmost cell of the tape—and the control unit assumes the initial/starting state. The machine then goes through its computation consisting of a sequence of steps. A computation may continue indefinitely, or it may terminate after some finite number of moves, usually a function of the contents and size of its input. If it does terminate, the symbols remaining on the tape are interpreted as the answer/outcome of the computation.

As described above, a Turing machine control unit is considered to be a set of rules for processing symbols on the tape. We should note that variations of a Turing machine (e.g., with multiple tapes, tracks, two-way infinite tape, etc.) are computationally no more powerful than the standard model described here. Obviously, an actual Turing machine cannot be built physically, since it would require an unlimited amount of tape. It is a formal model or an abstract specification for computation. Turing viewed numerical symbols essentially the same as nonnumerical symbols. That is, he considered numerical computation to be the same kind of work as, for example, playing chess or recognizing patterns.

Turing claimed that any intelligent work by a human brain can be simulated by a Turing machine. Therefore, we may consider that, in a certain sense, the work by Turing began the field of computer science known as artificial intelligence (AI). The famous *Turing test*, in which a human must determine if he or she is communicating with another human being or with a computer, is often used as an evaluation tool in AI.

A *Turing computable function* does not necessarily have to be a total function. In fact, for some inputs, a Turing machine may never terminate its computation (i.e., it may diverge). Besides solving problems, a Turing machine may also be viewed as a procedure for computing a function. We can compute a function according to the specifications of a Turing machine if we have enough memory and time. The specification of a Turing machine is also called the description of a Turing machine. If a Turing machine terminates its computation for every input taken from the input domain under consideration, the Turing machine is viewed as an *effective procedure* or an *algorithm* for computing that function.

A function is called a *total recursive function* (or simply a *recursive function*) if it is computed by a Turing machine that halts with an answer for every input. A function is called a *partially recursive function* if it is computed by a Turing machine, which does not necessarily always halt. Clearly, the set of recursive functions is properly included in the set of partially recursive functions. Interestingly, both sets have the same cardinality; i.e., they are countable (enumerable). This is not inconsistent; observe that the set of even numbers has the same cardinality as the set of natural numbers, although *N* contains even numbers as a proper subset.

A Turing machine may be viewed as computing a function from integers to integers. If a function has k arguments, k integers are initially placed on the tape in some appropriate form. For example, (2, 3, 2) may be represented by the string 001000100 (i.e., two 0s, three 0s, and two 0s, separated by 1s). Since there is a one-to-one correspondence between

N and the set of *k*-tuples of integers, any *k*-variable function may, without loss of generality, be considered a single-variable function.

For example, consider the problem of finding the greatest common divisor of a given positive integer pair. Since we know how to solve the problem (e.g., using Euclid's algorithm), this corresponds to computing a totally recursive function from the set of pairs of positive integers to the set of positive integers. The domain of the function is the set of pairs of positive integers, and its range is the set of positive integers.

Alternatively, computation of a partial (or total) recursive function may be viewed as an acceptance (or recognition) of a set of strings, i.e., a language. Consider the function $f(x) = x^2$, and its corresponding infinite language of strings in binary notation {0, 1, 100, 1001, ...}. The determination of whether an input string is a member of the language (the set of squares) is equivalent to the computation of the *characteristic function* of this set, and it is solving the "perfect square" problem. One can say that, computationally, the concepts of problem solving, recursive function-computation, and language-recognition are, in a sense, equivalent.

In addition to the formal definition of computability, Turing also presented a novel definition of a universal machine. This is now called *the universal Turing machine*. Such a machine can perform the work of any other Turing machine, provided that a description, or an index, of the other Turing machine is given to it. The universal Turing machine is like a general purpose computer. It can compute any computable function, as long as it is provided with an index (or a program) for computing the function.

In actual practice, we do not construct a dedicated computer for each computable function. Rather, we usually build a general purpose computer that computes the function when given a program for the function together with the argument values. A Turing machine corresponds to a program for computing a function. On the other hand, the universal Turing machine is an abstract model of a general purpose computer.

16.4 CHURCH–TURING'S THESIS

Turing submitted his paper on computability for publication on May 28, 1936, but just after that he learned of two papers by Alonzo Church (1903–1995) also published in 1936, and noticed that his definition of computability was equivalent to Church's notion of *effective calculability* [1, 2]. Turing therefore added an appendix to his paper, dated August 28, 1936, in which he mentioned the equivalence of the two definitions. The paper was published at the end of 1936 in the *Proceedings of the London Mathematical Society* [1].

From September 1936, Turing began a 2-year residency as a graduate student at Princeton University, where he completed the requirements for his doctorate with Alonzo Church as his thesis advisor. The activities of mathematicians and logicians such as Turing, Gödel, Church, Stephen Kleene (1909–1994), Emil Post (1897–1954), and others also gave rise to a wide variety of formalisms for the term *algorithm*, each endeavoring to describe the intuitive notion of an algorithm, effective procedure, or computability. All of these formalisms have been mathematically proven to be equivalent.

Church's thesis states, "It is believed that there are no functions that can be defined by humans, whose calculation can be described by any well defined algorithm that people can be taught to perform, that cannot be computed by Turing machines. The Turing machine is believed to be the ultimate calculating mechanism." A shorter version of Church's thesis simply states that "any computable problem can be computed by a Turing machine" [4–6].

The assumption that the intuitive notion of a computable function can be identical with the class of functions computable by Turing machines is now known as *Church–Turing's thesis*. We cannot hope to prove Church–Turing's thesis as long as the notion of a computable function remains informal. We can, however, present evidence to show that it is reasonable. As described above, logicians and mathematicians devised other formalisms for computable functions, and these formalisms have all been proven to be equivalent to Turing's definition of computability. Church–Turing's thesis has been universally accepted as valid.

If a function is computable by a Turing machine that always halts, then there must be an effective procedure (i.e., an algorithm) for computing it. In essence, the concepts of a Turing machine and of an algorithm (or a computer program) are similar in the sense that if an algorithm or a program solves a problem, then there exists a Turing machine that solves the same problem. A Turing machine solves a problem if it always gives a correct answer (in a finite number of steps) to any given instance x of the problem.

That is, men and computers are capable of computing it by means of an algorithm. An algorithm can be described either in a natural language like English or Japanese, or in a programming language like C++ or Java. Either way, an algorithm or program is a finite sequence of symbols. We therefore have as many algorithms or computer programs as there are integers. Given the fact that the cardinality of the set of functions significantly exceeds that of N, we must agree that almost all functions are not computable.

Does this sound extremely pessimistic? In fact, the situation is not as bad as it may seem, for although this is true, we can nevertheless avoid the limits of computability for many practical purposes. We should instead just make a note of the existence of functions/problems that are not solvable or even computable. The unsolvability of the halting problem implies that no general purpose computer program can always decide whether a given computer program will eventually terminate for every input given. However, such programs that work for a limited class of computer programs actually exist.

The formalization of Turing computability and of the intuitive notion of an effective procedure has been widely recognized as the origin of computation theory. The work originated by Gödel, Turing, Church, and others is among the greatest intellectual achievements of the 20th century.

REFERENCES
1. A. M. Turing, On Computable Numbers with an Application to the Entscheidungsproblem, *Proceedings of the London Mathematical Society 2*, 42, 230–265, 1936.
2. A. Church, An Unsolvable Problem of Elementary Number Theory, *American Journal of Mathematics*, 58, 345–363, 1936.
3. K. Gödel, Über formal unentscheidbare Sätze der Principia Mathematica und verwandter Systeme I, *Monatshefte für Mathematik Physik*, 38, 173–198, 1931.
4. J. E. Hopcroft and J. D. Ullman, *Introduction to Automata Theory, Languages, Computation*, Addison-Wesley, Reading, MA, 1979.
5. F. D. Lewis, *Essentials of Theoretical Computer Science*, Lecture Notes, University of Kentucky, Lexington, 1996.
6. M. Sipser, *Introduction to the Theory of Computation* (2nd ed.), Course Technology, Boston, MA, 2006.
7. Wikipedia, Alan Turing, http://en.wikipedia/wiki/Alan_Turing.

CHAPTER 17

Cryptography from the Medieval to the Modern Ages

17.1 THE ARAB CRYPTANALYSTS

A key for the substitution cipher consists of a permutation of 26 letters in the case of the Roman alphabet. As described in Chapter 7, the *shift cipher* (i.e., the Caesar cipher) is a special case of the substitution cipher. The number of permutations of 26 letters is 26!, which is more than 4×10^{26}, a very large number. Even if we restrict them to a smaller class of possible permutations, the number of these permutations is still very large. For example, as described in Section 7.2, pairing letters in the secret writing given in the Kama Sutra (written almost 2000 years ago in India) has more than 7×10^{12} possible keys. These large numbers of possible keys mean that the substitution cipher is an excellent way of secret writing. In fact, the mono-alphabetic substitution cipher had been widely used for many centuries. Many ancient scholars considered the substitution cipher unbreakable. However, Islamic scholars found shortcuts to break the cryptotexts in the ninth century. Arab cryptanalysts used linguistic and statistical analysis for breaking the cryptotexts, instead of trying all possible keys [1, 2].

The golden age of Islamic civilization began around the middle of the eighth century from the Abbasid caliphate after the transfer of the capital from Damascus to Baghdad. The Abbasid caliphs were less interested in military power than their predecessors. They respected the value of knowledge and expended much energy to build a wealthy society. For about 400 years from the middle of the eighth century, the Islamic world became the intellectual center for science, arts, philosophy, medicine, and education. The Islamic government encouraged business and industry, and protected its state documents by using cryptography [4].

The House of Wisdom in Baghdad was a library and translation institute established in 815 by the Abbasid caliph Al-Ma'mum (reigned from 813 to 833). From the 9th to the middle of the 13th century, it was considered a major intellectual center where both Islamic and non-Islamic scholars worked together on all areas of knowledge. Many classic works were translated into Arabic, and later, in turn, into Hebrew and Latin. During this period the Abbasid caliphate gathered knowledge from ancient Egypt, Greece, Rome, China, India, Persia, North Africa, and Byzantine [4]. However, the House of Wisdom and other scholarly institutes in Baghdad were destroyed during the Mongol invasion of 1258.

In addition to employing secret writing, the Arab scholars studied *cryptanalysis* (i.e., attempted to break ciphers). They succeeded in finding an efficient method for breaking the monoalphabetical substitution using a frequency analysis of letters and statistical analysis. A basic cryptanalytic attack against monosubstitution systems began with a frequency count of letters by Arab scholars. The number of occurrences of each letter in the cryptotext can be a crucial clue toward breaking it. The letters of the English alphabet are ordered to their frequency in Table 17.1 [1]. For instance, the letter with the highest frequency in the cryptotext encrypted from an English plaintext is likely to be the substitution for E, and this likelihood grows with the length of the cryptotext.

Although it is not known who first realized that linguistic analysis could be useful for breaking ciphers, the earliest known description of the frequency of letters used in cryptanalysis was by an Arab scholar, Al-Kindi (801–873). He was known as the great philosopher of the Arabs

TABLE 17.1 Frequency Ratios of the English Letters

E	T	A	O	N	I	S	R	H	L	D	C	U
12.31	9.59	8.05	7.94	7.19	7.18	6.59	6.03	5.14	4.03	3.69	3.20	3.10 (%)
P	F	M	W	Y	B	G	V	K	Q	X	J	Z
2.29	2.28	2.25	2.03	1.88	1.62	1.61	0.93	0.52	0.20	0.20	0.10	0.09 (%)

in the ninth century, and the author of 290 books on medicine, astronomy, mathematics, linguistics, and cryptography [5]. He gave the first explanation of cryptanalysis to monoalphabetic substitution ciphers in his book *A Manuscript on Deciphering Cryptographic Messages* [5].

17.2 POLYALPHABETIC SUBSTITUTION CIPHERS

A cryptosystem is called monoalphabetic if the use of substitutes remains unaltered throughout the plaintext, whereas polyalphabetic substitute systems use different substitutions in different parts of the plaintext. Until the 16th century monoalphabetic substitutes had been sufficient to maintain document secrecy. However, the development of frequency analysis made monoalphabetic substitution ciphers insecure.

Consequently, cryptographers tried to devise a stronger cipher that could withstand cryptanalytic attacks. Although polyalphabetic substitution ciphers had not been practically used until the 17th century, their origin can be traced back to a 15th-century Italian Renaissance polymath, Leon Battista Alberti (1404–1472) [1, 2, 6].

In 1467, Alberti wrote an essay on a new form of cipher. He proposed a set of two or more different cipher alphabets (i.e., permutations of the alphabet), and suggested switching these cipher alphabets during the enciphering process. In his new cryptosystem, the same letter in the plaintext is not necessarily to be replaced by the same corresponding letter in the cryptotext. Alberti's great cryptographic idea was spread not only in Italy, but also to other countries in Europe. The German abbot Johannes Trithemius (1462–1516), the Italian scientist Giovanni Porta (1513–1615), and the French diplomat Blaise de Vigenere (1523–1596), along with others, further developed Alberti's idea. Trithemius introduced a *tableau* (a large table) of the polyalphabetic cipher, and Porta described a sophisticated version of multiple cipher alphabets in his book *De Furtivis Literarum Notis* (On Concealed Characters in Writing) in 1563 [2].

In 1549, Blaise de Vigenere became acquainted with the work of Alberti, Trithemius, and Porta when he was sent to Rome on a diplomatic mission. He examined their ideas, and developed them into a stronger polyalphabetic substitution cipher that is known as the *Vigenere cipher*. In the encryption by the Vigenere cipher, a table called the *Vigenere square* (Table 17.2.) is first prepared. The ith row of the table represents a cipher alphabet with an i-step cyclic shift of the plaintext alphabet ($0 \leq i \leq 25$). In the Vigenere cipher, a different row of the table is used to encrypt different

TABLE 17.2   The Viginere Square

Keyword:	J O H N J O H N J O H N J O H
Plaintext:	A T T A C K T H E T R O O P S
Cryptotext:	J H A N L Y A U N H Y B X D Z

0	A B C D E F G H I J K L M N O P Q R S T U V W X Y Z
1	B C D E F G H I J K L M N O P Q R S T U V W X Y Z A
2	C D E F G H I J K L M N O P Q R S T U V W X Y Z A B
3	D E F G H I J K L M N O P Q R S T U V W X Y Z A B C
4	E F G H I J K L M N O P Q R S T U V W X Y Z A B C D
5	F G H I J K L M N O P Q R S T U V W X Y Z A B C D E
6	G H I J K L M N O P Q R S T U V W X Y Z A B C D E F
7	H I J K L M N O P Q R S T U V W X Y Z A B C D E F G
8	I J K L M N O P Q R S T U V W X Y Z A B C D E F G H
9	J K L M N O P Q R S T U V W X Y Z A B C D E F G H I
10	K L M N O P Q R S T U V W X Y Z A B C D E F G H I J
11	L M N O P Q R S T U V W X Y Z A B C D E F G H I J K
12	M N O P Q R S T U V W X Y Z A B C D E F G H I J K L
13	N O P Q R S T U V W X Y Z A B C D E F G H I J K L M
14	O P Q R S T U V W X Y Z A B C D E F G H I J K L M N
15	P Q R S T U V W X Y Z A B C D E F G H I J K L M N O
16	Q R S T U V W X Y Z A B C D E F G H I J K L M N O P
17	R S T U V W X Y Z A B C D E F G H I J K L M N O P Q
18	S T U V W X Y Z A B C D E F G H I J K L M N O P Q R
19	T U V W X Y Z A B C D E F G H I J K L M N O P Q R S
20	U V W X Y Z A B C D E F G H I J K L M N O P Q R S T
21	V W X Y Z A B C D E F G H I J K L M N O P Q R S T U
22	W X Y Z A B C D E F G H I J K L M N O P Q R S T U V
23	X Y Z A B C D E F G H I J K L M N O P Q R S T U V W
24	Y Z A B C D E F G H I J K L M N O P Q R S T U V W X
25	Z A B C D E F G H I J K L M N O P Q R S T U V W X Y

letters of the plaintext. A sequence of cyclic shifts to be applied is usually expressed by a keyword, which is shared between the sender and the receiver. For example, for a keyword *John*, the 9th row, the 14th row, the 7th row, the 13th row, again the 9th row, the 14th row, the 7th row, and so on, in the Vigenere square, are used to encrypt the letters of a plaintext in this order. If the intended receiver knows the keyword, then he or she can correctly choose which row of the Vigenere square is used to decrypt each letter of the received cryptotext. Blaise de Vigenere published his description of the Vigenere cipher in his book *Traicte des Chiffres ou Secretes Manieres d'Escrire* (Treatise on Secret Writing) in 1586 [1, 3, 7].

Example 17.1

Table 17.2 shows the encryption of a plaintext, "Attack the troops," by the Vigenere cipher with the keyword *John*.

The advantage of the Vigenere cipher is that it is invulnerable to attack by frequency analysis. The number of possible keywords used in the Vigenere cipher is enormous. Therefore, a cryptanalyst would be unable to break the cryptotext by searching all possible keywords. The polyalphabetic Vigenere cipher was considered an unbreakable cryptosystem until the middle of the 19th century. Although the Viginere cipher was not practically used for about two centuries after its discovery, it was frequently used in the 18th century [2, 7].

Charles Babbage was successful in the cryptanalysis of the Vigenere cipher in the 1850s. Babbage was an English mathematician, best known for his programmable mechanical calculating machine, the *analytical engine* (see also Chapter 11). His cryptanalytic discovery was not recognized for a century because he never published it. It was found in his notes in the 20th century. Friedrich Wilhelm Kasiski (1805–1881), a retired Prussian officer, independently discovered a cryptanalysis of the Vigenere cipher. He first published a general method for attacking Vigenere ciphers in 1863. Kasiski's technique is almost the same as Babbage's discovery. The technique is now called the *Kasiski test* [2, 7].

17.3 HOMOPHONIC SUBSTITUTION CIPHERS

The Vigenere cipher was sufficiently secure until the middle of the 19th century. A polyalphabetical substitution cipher was much more complicated to use than a monoalphabetic substitution cipher. For this reason, the Vigenere cipher had not been widely used until the 18th century. The monoalphabetic substitution cipher was adequate for many applications during the Renaissance period; however, it was inadequate for serious applications such as military and government communications in Europe in the 17th century. Consequently, cryptographers searched for a suitable cipher that was stronger than a monoalphabetic substitution cipher, but simpler to use than a polyalphabetical cipher. The *homophonic substitution cipher* was devised as a good candidate for these needs [2, 7, 8].

In a homophonic substitution cipher, each letter in the plaintext can be replaced with a variety of substitutes, where the number of potential substitutes of a letter is proportional to the frequency of the letter. For

TABLE 17.3   A Homophonic Substitution Cipher

A	:	01	12	33	47	53	67	78	92			
B	:	48	81									
C	:	13	41	62								
D	:	09	03	45	79							
E	:	14	16	24	44	46	55	57	64	74	82	87 98
F	:	10	31									
G	:	06	25									
H	:	23	39	50	56	65	68					
I	:	32	70	73	83	88	93					
J	:	15										
K	:	04										
L	:	26	37	51	84							
M	:	22	27									
N	:	18	36	59	66	71	91					
O	:	00	05	07	54	72	90	99				
P	:	38	95									
Q	:	94										
R	:	29	35	40	42	77	80					
S	:	11	19	58	76	86	96					
T	:	17	20	30	43	49	69	75	85	97		
U	:	08	61	63								
V	:	52										
W	:	60	89									
X	:	28										
Y	:	21	34									
Z	:	02										

Source: Adapted from Simon Singh, *The Code Book: The Science of Secrecy from Ancient Egypt to Quantum Cryptography,* Anchor Books, New York, 1999.

example, if we prepare 100 substitutes altogether, and if the frequency of letter β is approximately α_β% in plaintexts, we might choose one of the α_β substitutes for letter β in a cryptotext [2, 8]. An example of a homophonic substitution cipher is shown in Table 17.3. In the next example, we use the homophonic substitution given in Table 17.3.

Example 17.2

A plaintext, "Let us meet tonight in the park," may be encrypted to 51 82 69 08 86 22 44 16 97 49 90 18 70 25 65 75 32 66 17 39 98 95 78 29 04. Note that this is not a unique cryptotext for the plaintext.

For example, 37 64 75 61 11 27 14 98 30 20 07 91 88 25 23 49 32 66 43 65 55 38 33 77 04 is also a cryptotext that can be encrypted from the same plaintext.

In the 17th century, cryptographers tried to increase the strength of monoalphabetic substitution by incorporating the technique of the homophonic substitution ciphers. One such well-known cipher is the Great Cipher of Louis XIV. The Great Cipher was developed by Antoine Rossignol (1600–1682) and his son, and used to encrypt the king's important messages and records. The Rossignol family served the French crown as cryptographers. Modified forms of the Great Cipher were used by the French Army until the beginning of the 19th century [2, 9].

17.4 ENIGMA MACHINE

The earliest cryptographic machine is the cipher disk, invented in the 15th century by Italian polymath Leon Battista Alberti, one of the founders of the polyalphabetic cipher. Since then various cipher disks were used over the next five centuries. The cipher disk is a kind of a scrambler that takes each plaintext letter and transforms it into another letter or another letter sequence [2].

In 1918, at the end of World War I, German engineer Arthur Scherbius (1878–1929) applied for a patent for a cipher machine that can be considered an electromechanic version of Alberti's cipher disks. He and his friend founded a company to develop the cipher machine called *Enigma*, which became the most sophisticated cryptomachine. The first model of Enigma and its variants were used commercially from the 1920s. By 1925, Scherbius began mass-producing Enigma machines, and these Enigmas were adopted by German Nazi military and government services. The German military bought probably over 30,000 Enigmas. World War II began on September 1, 1939, with the invasion of Poland by Germany and subsequent declarations of war against Germany by the United Kingdom and France. The German military had obtained the most secure cryptosystems in the world before and during World War II [2, 10].

The Enigma machine is a combined electromechanical cryptosystem. It consists of a keyboard, a display board, a plug board, a set of rotating disks called rotors or scramblers, stepping components, and a reflector. The keyboard is used to input each plaintext letter. The display board consists of various lamps for indicating the cipher letters. The rotors form the major part of the Enigma machine and act as letter scramblers. Each rotor was a disk approximately 10 cm in diameter, made of hard rubber or Bakelite, with brass spring-loaded pins on one side arranged in a circle, and a circular electrical contact on the other side. The pins and contacts represent letters. The rotors were arranged along a spindle. The stepping components were used to turn one or more steps of the rotors with each key press. Different letter-letter substitutions were made by the movement of the rotors. The plug board was located between the keyboard and the first rotor. The Enigma user could insert cables in the machine in a certain way. Cable insertion caused some letters to be swapped before they entered the first rotor. The reflector was connected to the output of the last rotor so that the route of the letter stream could be altered. The use of the reflector in the Enigma machine started in 1926. The Enigma machine was contained in a compact box (size, 28 × 34 × 15 cm; weight, 12 kg) [2, 10].

The German Navy was first to adopt Enigma among the German military in 1926. The keyboard and display board contained 29 letters (A to Z and three German letters). Three rotors were chosen from a set of five rotors, and the reflector could be inserted in one of four different positions. The German Army also adopted the Enigma in 1928. A new version, called the *Wehrmacht Enigma*, was introduced in 1930. The new version was used extensively by the German military and other government organizations before and during World War II [2, 10].

17.5 BREAKING ENIGMA CODES

In the 1930s, Polish cryptanalysts were ahead of other countries in inventing various techniques for breaking the Enigma cipher. The Polish government formed the *Biuro Szyfrów* (the Cipher Bureau) in 1929–1930, and invited 20 mathematicians from Poznań University to join the bureau. The cryptanalysis developed by the Cipher Bureau in Poland on the early versions of Enigma was quite successful. Before World War II, Polish cryptanalysts had already designed an electromechanical machine, called the *Bomba*, to test the Enigma rotor setting. However, at the end of 1938 the German military modified its Enigma machines so the Polish Bomba was no longer capable of breaking the Enigma cipher [2, 10]. Five weeks before the German invasion of Poland, the Polish Cipher Bureau transferred two spare Enigma replicas and their techniques for breaking Enigma ciphers to the British and French governments.

The British government recruited a large number of mathematicians, scientists, college students, and young graduates as code breakers. They were sent to Bletchley Park, Buckinghamshire, the Government Code and Cypher School (60 km northwest of London). It was a newly established code-breaking organization. Initially, Bletchley Park had a staff of about 200, but within 5 years the number of workers there grew to about 7000. The well-known mathematician Max Newman was invited to Bletchley Park as one of the chief cryptanalysts [2, 12].

As described in Chapter 16, Alan M. Turing was one of the most influential mathematicians in the 20th century. He is known as the inventor of the abstract computing models, the so-called Turing machines. In 1939, the Government Code and Cypher School invited Turing to become a cryptanalyst at Bletchley Park. Turing's job at Bletchley Park was to find an effective way to break Enigma cryptotexts, even if the German military avoided using the same message key repeatedly. After investigating a large number of decrypted Enigma messages at Bletchley Park, Turing reached some ideas for finding the weakness of Enigma ciphertexts. Turing started designing electrical circuits that would remove the effect of the Enigma plug board. In 1939, Turing completed his design of the device, and it was called the *Bombe*. More than 200 Bombes were built by the British Tabulating Machine Company at Letchworth during World War II, but all of them were destroyed after the war [2, 12, 13].

After the end of World War II Turing worked at the National Physical Laboratory, where he designed a stored-program computer, the Automatic Computing Engine (ACE) (see also Chapter 18). His sexual orientation resulted in a criminal prosecution in 1952 since homosexual acts were illegal in the United Kingdom at that time. Over the next 2 years he became severely depressed. He dipped an apple in cyanide and took several bites on June 7, 1954. The greatest mathematician of the 20th century and the Father of Computer Science committed suicide just a few weeks before his 42nd birthday. On September 10, 2009, British Prime Minister Gordon Brown made an official public apology on behalf of the British government for their treatment of Turing for several years after World War II [14].

17.6 LORENZ CIPHER

The Lorenz SZ40 and SZ42 were German cipher machines used during World War II. Those cipher machines were made by Lorenz Company, and were even more complex than the Enigma. Bletchley Park code breakers called the Lorenz machine *Tunny*, and the cipher messages by Tunny were called *Fish*. While the Enigma was mainly used in field units, the Lorenz was used exclusively for the most important messages between the German

Army field marshals and their Central High Command in Berlin. It was not a portable device. The size of the machine itself was 51 × 46 × 46 cm, but it could support the heavy teletypewriter and attendant fixed circuits [2, 11].

The Lorenz used the international telephone code in which each letter of the alphabet was represented by a series of five electrical impulses. Messages were encrypted by the *exclusive OR* operation (the bitwise addition of modulo 2) of five pseudorandom bits and the plaintext. The obscuring letters were generated by 12 rotors, another 5 of which followed a regular pattern, and another 5 of which followed a pattern dictated by two pin wheels. Cracking Fish (cryptotext) relied on determining the starting configuration of the Lorenz machine's rotors [2, 14].

A British cryptanalyst, John Tilman, broke Fish messages at Bletchley Park in 1941 using hand methods that relied on statistical analysis, but the Germans had introduced complications that made it almost impossible to break Lorenz ciphers by hand only. British mathematician Max Newman was assigned to the Research Section at Bletchley Park to work as a cryptanalyst of Lorenz ciphers. He proposed that the code-breaking process could be mechanized. In December 1942, he was assigned to build suitable machines for that purpose. The first machine designed to break the Lorenz cipher was built at the Post Office Research Department at Dollis Hill and was called *Heath Robinson*. Although the Heath Robinson worked well enough, it was rather slow [2, 11, 15].

Newman asked for the help of Tommy Flowers, a post office electrical engineer at Dollis Hill in London. Flowers built a much faster and more reliable machine called *Colossus* that used 1500 vacuum tubes. The first Colossus arrived at Bletchley Park in December 1943. This was the first electronic digital information processing machine in the world. The Lorenz cryptotexts could be deciphered by carrying out complex statistical analysis on intercepted messages. Colossus could read paper tape at 5000 characters per second, and the paper tape moved at 30 miles per hour [2, 11, 16] (see also Chapter 18).

REFERENCES

1. Arto Salomaa, *Public-Key Cryptography*, Springer-Verlag, Berlin, 1990.
2. Simon Singh, *The Code Book: The Science of Secrecy from Ancient Egypt to Quantum Cryptography*, Anchor Books, New York, 1999.
3. Douglas R. Stinson, *Cryptography: Theory and Practice*, CRC Press, New York, 1995.
4. Wikipedia, Abbasid Caliphate, http://en.wikipedia.org/wiki/Abbasid_Caliphate.
5. Wikipedia, Al-Kindi, http://en.wikipedia.org/wiki/Al-Kindi.

6. Wikipedia, Leon Battista Alberti, http://en.wikipedia.org/wiki/Leon_Battista_Alberti.

7. Wikipedia, Vigenere Cipher, http://en.wikipedia.org/wiki/Vigenere_cipher.

8. Wikipedia, Substitution Cipher, http://en.wikipedia.org/wiki/Substitution_cipher.

9. Wikipedia, Great Cipher, http://en.wikipedia.org/wiki/Great_Cipher.

10. Wikipedia, Enigma Machine, http://en.wikipedia.org/wiki/Enigma_machine.

11. Wikipedia, Lorenz Cipher, http://en.wikipedia.org/wiki/Lorenz_cipher.

12. Wikipedia, Bletchley Park, http://en.wikipedia.org/wiki/Bletchley_Park.

13. Wikipedia, Bombe, http://en.wikipedia.org/wiki/Bombe.

14. Wikipedia, Alan Turing, http://en.wikipedia.org/wiki/Alan_Turing.

15. Wikipedia, Heath Robinson (code breaking machine), http://en.wikipedia.org/wiki/Heath_Robinson.

16. Wikipedia, Colossus Computer, http://en.wikipedia.org/wiki/Colossus_computer.

CHAPTER 18

Electronic Computers

Determining when, where, and by whom the first electronic digital computer was invented is a daunting task as described by A. R. Burks, the author of *Who Invented the Computer?* [1]. The history of this revolutionary development is surrounded by the events of World War II and clouded by the secrecy required by advanced-level military research. The race to decode enemy communications, to develop reliable ballistic tables, and to build the first rockets and warheads brought together some of the most brilliant mathematicians and engineers of the 20th century. Their work, sometimes in collaboration, while at other times independently, led in fits and starts to the invention of electronic digital computers, the forerunners of the computers we enjoy today. What follows is a brief discussion of some of their most significant and hard-won contributions.

18.1 THE ABC COMPUTER

Born in Hamilton, New York, John Vincent Atanasoff (1903–1995) was an American physicist and inventor. In 1925, Atanasoff received his B.Sc. degree in electrical engineering from the University of Florida, and in 1926 he earned an M.Sc. degree in mathematics at Iowa State College (now Iowa State University). Atanasoff received a Ph.D. in theoretical physics from the University of Wisconsin, Madison, in 1930, after which he obtained an academic position at Iowa State College in mathematics and physics [2].

In 1939, Professor Atanasoff and his graduate assistant, Clifford Berry (1918–1963), began building the world's first electronic-digital computer at Iowa State College, working on the project for the next few years. The

Atanasoff–Berry Computer (ABC) contained various innovations in computing, such as binary number systems, digital circuits using vacuum tubes, regenerative memory called *capacitors*, and a separation of memory and computing functions [3, 6]. The capacitors were built in a rotating drum that held electric charges representing the memory for binary numbers. The prototype of their computer won them a grant of $850 to build a full-scale model. The final product weighed 318 kg and had more than 300 vacuum tubes. It could perform one fundamental operation per 15 seconds [6].

The electronic part of the ABC was successful, but the reliability of its binary card reader was unsatisfactory. The project was discontinued when Atanasoff left Iowa State College, and the ABC was dismantled.

18.2 THE Z3 COMPUTER

A German engineer and computer pioneer, Konrad Zuse's (1910–1995) great achievement was the invention of one of the world's first electrical computers, the Z3, which became operational in 1941 [8].

The basic components of the Z3 were small, electrically driven, mechanical switches called *relays*, making Z3 an electromechanical digital computing machine. Several similar digital computing machines were built before and during World War II by Howard Aiken (1900–1973) at Harvard University, George Stibitz (1904–1995) at Bell Telephone Laboratories, and Alan M. Turing at Bletchley Park. Among them, Zuse received the honor of having built the first working general purpose program-controlled digital computer (the Z3). A program-controlled computer, as opposed to a stored-program computer, is set up for a task by reconfiguring the wires (e.g., by means of plugs) [5, 7].

Zuse designed the Z1 between 1935 and 1936 and built it between 1936 and 1938. It was totally mechanical, but unreliable. Zuse decided to base his next design for the Z2 on the use of relays. The Z2 was completed in 1939 and demonstrated to *Die Deutsche Versuchsanstalt für Luftfahrt* (the German Laboratory for Aviation) in 1940. Further improving the Z2 computer, he went on to build the Z3 in 1941. His work was a top secret project of the German government. The Z3 was faster and far more reliable than either the Z1 or the Z2 and was built with about 2000 relays, implemented on a 22-bit word length with a clock frequency of 5 to 10 (Hz). Zuse asked the German government for funding to replace the relays with fully electronic switches, but it was denied during World War II, since such development was not considered urgent during World War II [5, 7, 9].

The success of Zuse's Z3 is often attributed to its use of the binary number system. However, performing arithmetic with a binary number system in a calculating device was invented roughly three centuries earlier by Gottfried Leibniz. George Boole later used it to develop his Boolean algebra. In 1937, Claude Shannon (1916–2001) introduced the idea of mapping Boolean algebra to electrical relays in his work on digital circuit design. It seemed that Zuse was not aware of Shannon's work and developed digital circuits independently.

The original Z3 was destroyed in 1943 during an Allied bombardment of Berlin. Zuse's coworker Helmut Schreyer built an electrodigital prototype experimental model of a computer using 100 vacuum tubes in 1942, but it was also lost at the end of World War II. A fully functioning replica was built in the 1960s by Zuse's company. It is exhibited in the Deutsche Museum.

18.3 THE COLOSSUS COMPUTER

In 1939, the day after the war broke out, Alan M. Turing enlisted full-time at the British Government Code and Cypher School at Bletchley Park, 50 miles northwest of London in Buckinghamshire. He was a member of the group of able mathematicians drafted into the military's code-breaking operations [16].

Tommy Flowers (1905–1998) was an English engineer born in London. He took evening classes at the University of London to earn a degree in electrical engineering. In 1929, he joined the telecommunication branch of the General Post Office. He was one of the earliest extensive users of vacuum tubes for digital data processing. In 1934 Flowers designed electronic equipment for controlling the connection between telephone exchanges. This device went into operation in 1939. From 1938 to 1939, Flowers worked on an experimental electronic digital data processing system with a high-speed data store.

Turing wanted Flowers to build a decoder for the relay-based machine, called the *Bombe*, which Turing had developed to help decrypt the Enigma codes. The British Government Code and Cypher School was successfully deciphering German radio communications by means of the Enigma system, and by early 1942 about 39,000 intercepted messages were being decoded each month by using these electromechanical Bombe machines [16].

Turing introduced Flowers to Max Newman, who headed the team for breaking the German code cipher generated by a teletypewriter coding machine, the *Lorenz*, one of the German *Geheimschreiber* (secret writer) systems, which the British called *Tunny*, a far more complex coding system than Enigma. The need to decipher Tunny codes as rapidly as possible led Max Newman to propose in 1942 that the key parts of the decryption process be automated by means of high-speed electronic processing devices. The first machine designed and built to Newman's specification, known as the *Heath Robinson*, was relay based and used vacuum tubes in part. The Heath Robinson was installed in 1943, but was unreliable and slow. However, it proved that Newman's idea for breaking the Lorenz code was worth the effort [14, 16].

Flowers recommended building an all-electronic machine instead. Obtaining full backing for his project, Flowers built the first large-scale programmable electronic digital Tunny code-breaking computer, called the *Colossus I*, at the Research Station in the General Post Office at Dollis Hill in northwest London. He delivered it to Bletchley Park in 1943, and by the end of the war there were 10 Colossi working at Bletchley Park where they were used by British code breakers to help decipher encrypted German messages during World War II. The Colossus I contained approximately 1600 vacuum tubes, though each of the subsequent machines had approximately 2400 vacuum tubes (Figure 18.1). The Colossus lacked two important features of modern computers: First, it had no internally stored programs. To program it for a new task, the operator had to reconfigure the machine's physical wiring, using plugs and switches. Second, the

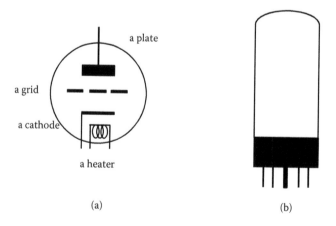

Glass Tube (1940–1950)

FIGURE 18.1 An example of a vacuum tube. (a) The structure of a triode; (b) the shape of a glass tube (1940–1950).

Colossus was not a general purpose machine since it was designed for a specific cryptanalytic task [9, 12, 16].

For security reasons, most of the Colossi were destroyed after the end of the war, but two Colossi were retained by the Government Code and Cypher School, renamed the Government Communication Head Quarters (GCHQ), even after the end of the war [4]. The last Colossus was believed to have stopped running in 1960. Until the 1970s, few had known that the Colossus was used successfully for code breaking during World War II. Irving John (Jack) Good (1916–2009) and Donald Michie (1923–2007) published their notes on the Colossus in 1970 and 1975, respectively. During the war, Good worked at Bletchley Park and contributed to the development of the Colossus, and Michie also had the experience of working there. A replica of the Colossus computer was completed in 2007, and is displayed in the National Museum of Computing at Bletchley Park. Now the Colossus is widely believed to be the world's first electronic digital programmable computer [9].

18.4 THE ENIAC COMPUTER

John Mauchly (1907–1980) was an American physicist who, along with John Presper Eckert (1919–1995), designed the Electronic Numerical Integrator and Computer (ENIAC). Mauchly was born in Cincinnati, Ohio. He completed his Ph.D. in physics at Johns Hopkins University in 1932. In 1941, Mauchly took a course in wartime electronics at the Moore School of Electrical Engineering, University of Pennsylvania, where he met a Moore School graduate student, Eckert. The Moore School was a center for wartime computing. The critical problem at the Moore School was the ballistic calculation that the U.S. military was developing for the war effort [10, 11].

The U.S. military needed a calculating machine for preparing artillery firing tables. The tables would be used for different weapons under varied conditions so that the target would be set accurately. The U.S. Army's Ballistics Research Laboratory heard about Mauchly's research in the Moore School at the University of Pennsylvania. Mauchly had previously created several calculating machines, some with small electric motors inside. In 1942, Mauchly had begun designing a better calculating machine that would use vacuum tubes to speed up calculations.

ENIAC, designed to calculate and construct artillery firing tables for the Ballistic Research Laboratory, was financed by the U.S. Army during World War II. The construction contract was signed in 1943, and began in secret at the Moore School of Electrical Engineering. Mauchly and Eckert were the chief consultant and the chief engineer of the project, respectively. The machine took about a year to design, and another 18 months and $500,000 to build, with completion coming in November 1945. The ENIAC continued to be used to perform various calculations for advanced military research projects after the war was over [1, 6, 10, 15].

The size of the ENIAC was remarkable. It contained 17,468 vacuum tubes, 7,200 crystal diodes, 70,000 resistors, 10,000 capacitors, 1,500 relays, 6,000 manual switches, and 5 million soldered joints. The machine covered 1,800 square feet (167 square meters) of floor space, weighed 30 tons, and consumed 160 kilowatts of electrical power. Input was provided with an IBM card reader, while an IBM card punch was used for output. The ENIAC used 10-position ring counters to store digits. Arithmetic was performed by counting pulses with the ring counters and generating carry pulses if the counter wrapped around. It had 20 ten-digit signed accumulators that used 10's complement representation and could perform 5,000 simple addition or subtraction operations per second. It could perform 357 multiplications or 38 divisions per second [6, 9, 10].

The use of vacuum tubes increased its speed, but the ENIAC was not easy to reprogram. In fact, one significant problem with the ENIAC was that it was extremely difficult to program. It had to be hardwired afresh for solving each new problem by reprogramming it with plug panels. These changes took technicians several days of tedious manual rewiring. Another serious problem was its reliability. Vacuum tubes burned out frequently since special high-reliability tubes were not available until 1948. Most vacuum tube failures occurred during the warm-up and cool-down periods, when the tube heaters and cathodes overheated. According to an interview with Eckert, a tube failure occurred every 2 days, but the technicians could locate the problem within 5 minutes. In 1954, the longest continuous period of operation without a failure was 116 hours. In 1948, John von Neumann (1903–1957) made several modifications to the ENIAC [10].

In 1946, the Moore School decided to change its patent policy in order to gain commercial rights to any future and past computer development there. Eckert and Mauchly decided this was unacceptable, and they resigned their positions. In 1947, they formed the Eckert–Mauchly Computer Corporation. In 1949, their company launched the Binary Automatic Computer (BINAC) that used magnetic tape to store data [1].

In 1950, Remington Rand Corporation bought Eckert–Mauchly Computer Corporation, changing the name to the UNIVAC Division of Remington Rand. Their research resulted in the Universal Automatic Computer (UNIVAC), an important forerunner of modern computers. In 1955, Remington Rand merged with Sperry Corporation, forming Sperry–Rand. Eckert remained with the company as an executive, staying on as it later merged with Burroughs Corporation to become Unisys.

The ENIAC retired when power was finally shut off on October 2, 1955 [10].

18.5 VON NEUMANN ARCHITECTURE FOR COMPUTERS

John von Neumann was born and educated in Budapest, Hungary. He received his Ph.D. in mathematics (as well as in experimental physics and chemistry) from Pater University in Budapest in 1928. In 1930, von Neumann was invited to Princeton University, and then offered a position at the Institute for Advanced Study there in 1933. He was highly regarded in the fields of set theory, algebra, quantum physics, and computing. Von Neumann retained the position of a professor in mathematics at the Institute for Advanced Study for the remainder of his life [17].

From 1936 to 1938, Alan M. Turing was a visitor in the Institute for Advanced Study at Princeton University. He completed his Ph.D. dissertation under Alonzo Church's supervision. We could imagine that von Neumann and Turing met in Princeton during this period and that von Neumann knew of Turing's ideas about computability and the universal Turing machine. However, it is unknown how much they discussed computers and related subjects at Princeton. Turing returned to Cambridge, England, and a year later he was involved in war work at Bletchley Park. To what extent von Neumann used Turing's ideas for his design of computers 10 years later remains unknown.

During World War II, von Neumann was deeply involved as a consultant to the armed forces, participating also in the development of the atomic bomb. Toward the end of World War II, von Neumann took part in several national committees, serving as a contact person between groups of scientists and government organizations. He worked as a consultant with the Los Alamos National Laboratory, the Manhattan Project, and as an adviser to the engineer group, building the ENIAC at the Moore School of Electrical Engineering [10, 17].

Von Neumann joined Electronic Discrete Variable Automatic Computer (EDVAC), a project that began in 1944 as the successor of ENIAC at the Moore School, University of Pennsylvania. Von Neumann's ideas about the structure of a computer eventually became the fundamental organization of the modern computer, now known as *von Neumann architecture*.

The basic elements of the EDVAC were based on the stored-program concept. His work as a project consultant included preparing the "First Draft of a Report on the EDVAC," written in the spring of 1945. The draft, distributed to the staff of the Moore School of Electrical Engineering, presented the stored-program concept and the overall structure of a computer system.

The report organized the computer system into four main parts: the central arithmetic unit, the central control unit, the memory, and the input/output devices. The central arithmetic unit carried out the four basic arithmetic operations and some higher arithmetic functions, such as roots, logarithms, trigonometric functions, and their inverses. The control unit controlled the proper sequence of operations and made the individual units act together to carry out the specific programmed task. The memory stored both numerical data and numerically coded instructions, and the input/output devices served as the user's computer interface. It described how these four parts communicate with each other to process information. However, the specific materials and design of the implementation of each unit were not recorded in the report. Although "First Draft of a Report on the EDVAC" was authored solely by von Neumann, the basic idea about the stored program was derived from his discussions with Eckert, Mauchly, and others [18].

Von Neumann's contributions to computer design were great, but he was less interested in patents and patent law. Most of the information about his innovations, such as his "First Draft of a Report on the EDVAC," was widely distributed. He was happy to share his thoughts and theories with anyone who was interested in computer design. Von Neumann left the EDVAC project in 1946, and then returned to Princeton University, where he was involved in the later Institute for Advanced Study (IAS) computer. His basic architectural design can be easily recognized even in the most advanced computers of today. The IAS computer also had a strong influence on the IBM 701 built in 1952, which was the first mass-produced electronic stored-program computer produced by International Business Machines (IBM).

18.6 OTHER NOTABLE EARLY ELECTRONIC COMPUTERS

18.6.1 National Physics Laboratory and the ACE

At the end of World War II John Ronald Womersley (1907–1958) was appointed superintendent of the Mathematics Division of the National

Physical Laboratory in England, where he coined the name *Automatic Computing Engine* (ACE) for the early electronic computer developed there. Alan Turing was also asked to use his theories and experience for the ACE project [13]. Although Turing was not directly involved in the hardware development of the Colossus project, he saw the potential of the electronic computer to realize a computing machine that could carry out processes previously assumed possible only by the human brain. In 1946 Turing presented a detailed paper to the National Physical Laboratory Executive Committee, giving a reasonably complete design of a stored-program computer [13]. Turing's report on the ACE included detailed logical circuit diagrams and a cost estimate of 11,200 pounds.

Unlike the EDVAC, the ACE implemented subroutine calls, and an additional departure from the EDVAC was the use of Abbreviated Computer Instructions, an early form of programming language. The first version of the ACE was a smaller version of Turing's original design. The Pilot ACE had 1,450 vacuum tubes, and used mercury delay lines for its main memory. Each of the 12 delay lines could store 32 instructions or data words of 32 bits. Turing resigned from the National Physical Laboratory in 1948 and moved to Manchester. The Pilot ACE ran its first program on May 10, 1950. With an operating speed of 1 MHz, the Pilot ACE was for some time the fastest computer in the world [13].

18.6.2 The MARK 1 at Manchester University

In September 1945, Max Newman was appointed professor in the mathematics department at the University of Manchester. The earliest general purpose stored-program electronic computer was built in Newman's Computing Machine Laboratory. The *Manchester Baby*, as it became known, was constructed by engineers Frederic Calland Williams (1911–1977) and Tom Kilburn (1921–2001), and performed its first calculation in June 1948. That year, Turing joined the mathematics department at the University of Manchester as deputy director of the Royal Society Computing Machine Laboratory. He designed an enlarged version of the Manchester Baby that became the world's first commercially available computer, the Mark I. The first Manchester Mark I was completed in 1951, and installed at the University of Manchester. About 10 Manchester Mark I computers were sold in Britain, Canada, Holland, and Italy.

Turing used the Manchester Mark I to investigate prime numbers in collaboration with Newman. Meanwhile, he continued his theoretical work and in 1950 published another famous paper, "Computing Machinery and

Intelligence," in which he asked the important question "Can computers think?" anticipating the subject of artificial intelligence. Turing's main contribution to the Manchester Mark I project was providing the early software requirements for computers, and writing the first programming manual.

By 1951, Newman and Turing had withdrawn from active involvement in the Manchester Mark I project and subsequent computer development. However, Turing was still a keen user of the computer as a tool for his research interests, and was always ready to help programmers of the Manchester Mark I with their problems.

18.6.3 Electronic Delay Storage Automatic Calculator (EDSAC)

In 1949, the EDSAC was built at Cambridge University by Maurice Wilkes (1913–2010) and was in operation there until 1958. In 1950, a British computer company, J. Lyons and Co., revised the EDSAC, selling it in the commercial market as the Lyons Electronic Office (LEO) computer. Wilkes received the Turing Award in 1967 for the design and construction of the EDSAC.

18.6.4 Whirlwind I

Whirlwind was developed at MIT. It is the first computer operated in real time, using video displays for output. By 1947, Joy Forrester (1918–) and collaborator Robert Everett (1921–) completed the design of a high-speed stored program. It first went online in 1951.

18.6.5 Standards Eastern Automatic Computer (SEAC)

SEAC was an early electronic computer, built by the U.S. National Bureau of Standards. In 1950, it went into full production, making it the first fully functional stored-program electronic computer in the United States.

18.6.6 Standards Western Automatic Computer (SWAC)

SWAC was built in 1950 by the U.S. Bureau of Standards Western Division and Institute for Numerical Analysis, University of California, Los Angeles. It was designed by Harry Huskey (1916–).

REFERENCES

1. A. R. Burks, *Who Invented the Computer? The Legal Battle That Changed Computer History*, Prometheus Books, Amherst, New York, 1996.
2. Wikipedia, John Vincent Atanasoff, http://en.wikipedia.org/wiki/John_Vincent_Atanasoff.

3. Wikipedia, Atanasoff—Berry Computer, http://en.wikipedia.org/wiki/Atana soffAtanasoff%E2%89%93Berry_Computer.
4. Stanford Encyclopedia of Philosophy, The Modern History of Computing, http://plato.stanford.edu/entries/computing-history/.
5. Wikipedia, Z3 (computer), http://en.wikipedia.org/wiki/Z3_(computer).
6. Mary Bellis, Inventors of the Modern Computer, http://inventors.about.com/library/weekly/aa050898.html.
7. J. Alex, H. Flessner, W. Mons, K. Pauli, and H. Zuse, *Konrad Zuse: Der Vater des Computers*, Verlag Parzeller, Fulda, 2000.
8. Wikipedia, Konrad Zuse, http://en.wikipedia.org/wiki/Konrad_Zuse.
9. J. Palfreman and D. Swade, *The Dream Machine*, BBC Books, London, 1991.
10. Wikipedia, ENIAC, http://en.wikipedia.org/wiki/ENIAC.
11. Wikipedia, John Mauchly, http://en.wikipedia.org/wiki/John_Mauchly.
12. Wikipedia, Colossus Computer, http://en.wikipedia.org/wiki/Colossus_computer.
13. Wikipedia, Automatic Computing Engine, http://en.wikipedia.org/wiki/Automatic_Computing_Engine.
14. Wikipedia, Max Newman, http://en.wikipedia.org/wiki/Max_Newman.
15. M. Campbell-Kelly and W. Aspray, *Computer: A History of the Information Machine*, Basic Books/HarperCollins, New York, 1996.
16. S. Singh, *The Code Book*, Anchor Books, New York, 1999.
17. Wikipedia, John von Neumann, http://en.wikipedia.org/wiki/John_von_Neumann.
18. Wikipedia, EDVAC, http://en.wikipedia.org/wiki/EDVAC.

CHAPTER 19

Numerical Methods

After successfully conquering the simple arithmetical tasks of counting, addition, subtraction, multiplication, and division, scholars slowly began to address significantly more challenging mathematical problems. These started with the calculation of $\sqrt{2}$, π, areas and volumes of objects, and astronomy (orbit projections of heavenly bodies). Eventually, this led to the development of well-defined numerical methods and procedures, which were initially performed by hand, and later by a computer. Initially, fundamental problems in physics, in particular, motion, led to the development of many innovative numerical methods. This is now a well-established branch of mathematics.

19.1 NUMERICAL CALCULATION IN ANCIENT CIVILIZATIONS

When the ancient Greeks learned about the irrationality of $\sqrt{2}$, i.e., it is not expressible as a ratio of two integers and therefore cannot be measured using numbers known to them, Pythagoras (c. 569–475 BC) and his followers decided to keep this sensational discovery a *secret*. It could only be revealed to the initiated insiders, called the *mathematikoi* (the learners). Legend has it that the man who disclosed this secret was thrown overboard and drowned at sea. It is hypothesized that the unlucky fellow was actually Hippasus of Metapontum (ca. 500 BC). He is often credited with obtaining the first classical proof of $\sqrt{2}$'s irrationality. The proof is based on the *unique* factorization of any integer into primes and because the square of any fraction of the form p/q (where p and q are integers) features an even number of prime factors both in the numerator and in the denominator, which cannot cancel pairwise to yield a single prime, e.g., 2, in lowest terms. The value of $\sqrt{2}$ is sometimes referred to as the *Pythagorean constant*, which is somewhat ironic, especially given that it was Pythagoras himself who wanted to keep the irrationality of $\sqrt{2}$ a secret (see also Chapter 3).

The existence of the constant represented by π, i.e., the ratio of the diameter to the circumference of a circle, has been known and understood by scholars for several thousand years, almost from the beginning of man's recorded history. Ancient Greeks suspected that, just as was the case with $\sqrt{2}$, π cannot be represented in the form of *p/q*. The history of attempts to calculate the exact value of π spans for at least four millennia, starting with the ancient Babylonians, Egyptians, Chinese, Indians, and Europeans, including such notable scholars as Archimedes, Euclid, Euler, Fibonacci, Leibniz, and Newton.

The actual symbol π, which is probably the most famous among all of the transcendental numbers, was first introduced by William Jones (1675–1749) in 1706, probably because it is the first letter of the Greek word *perimetros*, from which the word *perimeter* is derived. Its usage was popularized by the Swiss mathematician, Leonhard Euler (1707–1783). Although many scholars suspected that it is not rational, the irrationality of π was formally proved only in 1761 by a Swiss mathematician, Johann Lambert (1728–1777) (see also Chapter 13).

In one of the earliest recorded accounts about π, the Babylonians (ca. 2000 BC) used an approximation of 3 + 1/8 = 3.125. Ancient Egyptians approximated the value of π to be about 3.16, whereas the Bible (Old Testament) simply used the whole integer 3. Archimedes of Syracuse obtained lower and upper bounds for π that were fairly accurate (two digits past the decimal point). A number of ever-improving approximations for π were obtained throughout the ages (see Table 19.1, but note that it is not exhaustive), and all of the calculations until 1946 were carried out by hand. These were usually obtained from approximations based on polygons with an ever-increasing number of sides, as shown in Figure 19.1, or other methods; e.g., Leibniz and Newton were able to apply some of their formulas from calculus (invented by them) to estimate the value of π.

Starting in the mid-1940s electronic computer technology allowed the value of π to be computed with, what now almost amounts to, arbitrary precision. These computations were facilitated by the ever-faster hardware

TABLE 19.1　Accuracy over Time Improvements (Measured in Digits) for Calculations of π

Hand Calculations (from 2000 BC to 1946)			Computer-Assisted Calculations (from 1947 to 2010)			
Individuals	Date	Digits	Programmers	Date	Digits	Computer
Babylonians	2000 BC	1	Ferguson	1947	710	Desk Calculator
Egyptians	2000 BC	1	Smith and Wrench Jr.	1949	1 120	Desk Calculator
Archimedes	250 BC	2	Reitwiesner et al.	1949	2 037	ENIAC
Ptolemy	150	3	Nicholson and Jeenel	1954	3 092	NORC
Liu Hui	263	5	Felton	1958	10 020	Pegasus
ZuChongzhi	480	7	Guilloud	1959	16 167	IBM 704
Al-Kashi	1429	14	Shanks and Wrench	1961	100 265	IBM 7090
Romanus	1593	15	Guilloud and Filliatre	1966	250 000	IBM 7030
van Ceulen	1615	35	Guilloud and Dichampt	1967	500 000	CDC 6600
Grienberger	1630	39	Guilloud and Bouyer	1973	1 001 250	CDC 7600
Sharp	1699	71	Kanada and Miyoshi	1981	2 000 036	FACOM M-200
Machin	1706	100	Tamura	1982	2 097 144	MELCOM 900II
De Lagny	1719	112	Tamura and Kanada	1982	8 388 576	Hitachi M-280H
Vega	1794	136	Gosper	1985	17 526 200	Symbolics 3670
Rutherford	1841	152	Bailey	1986	29 360 111	CRAY-2
Dahse	1844	200	Kanada et al.	1987	134 214 700	NEC SX-2
Clausen	1847	248	Kanada and Tamura	1988	1 073 741 799	Hitachi S-820/80
Lehmann	1853	261	Chudnovskys	1991	2 260 000 000	m-Zero
Rutherford	1853	440	Kanada and Takahashi	1999	206 158 430 000	Hitachi SR8000
Richter	1854	500	Kanada et al.	2002	1 241 100 000 000	Hit. SR8000/MP
Shanks	1874	527	Takahashi et al.	2009	2 576 980 377 524	T2K Open
Ferguson	1946	620	Yee and Kondo	2010	5 000 000 000 000	y-Cruncher

Note:　The data for this table are a selective compilation from over 30 sources, including books, scholarly papers, mass media, and the Internet; these will not be cited individually. Numerous inconsistencies in the dates and the accuracies obtained, especially in the pre-1947 computations, were resolved by using the "majority vote" approach. There was no averaging.

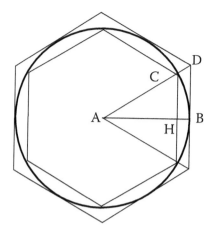

FIGURE 19.1   Inscribed and circumscribed polygons for computing π.

and the discovery of very advanced algorithms for performing the required high-precision floating-point arithmetic operations on a computer. A significantly condensed historical summary of π's calculations by computers from 1947 until 2010 is given in Table 19.1. For example, in 2009, Japanese T2K Open Supercomputer more than doubled the previous record by calculating 2,576,980,377,524 digits of π. This was followed with a 2010 calculation by Shigeru Kondo, who used Alexander Yee's program and his own home-built computer, called the *y-cruncher*, to calculate the first 5,000,000,000,000 digits of π.

From the scientific or engineering perspective, there is no need to calculate the value of π to more than, say, 1000 digits. Whereas the computations of the billions, trillions, or even quadrillions of π's digits are not necessary, some potential benefits, such as the testing of computers' hardware/software integrity, may be reaped from these modern-day computational marathons. A humorous side note: March 14 (3.14) has been designated as π-day. This also happens to be Albert Einstein's (1879–1955) birthday (Table 19.1).

19.2 NUMERICAL SOLUTION OF ALGEBRAIC EQUATIONS

Numerical analysis, which also predates the age of the modern electronic computer, is the study of algorithms that use numerical approximations to represent real numbers, and mathematical objects such as curves, surfaces, functions, and related phenomena. Much of numerical analysis is concerned with obtaining approximate solutions, while at the same time maintaining certain acceptable bounds on the computational errors. Just as important for these numerical methods is the determination of the rate of convergence, i.e., how fast is the answer actually produced. Numerical analysis has applications in all fields of engineering, physical sciences, economics, and many other areas of social sciences, and medicine.

For centuries mathematicians were preoccupied with the construction of numeric methods [1] that would allow them to solve problems that were practical (e.g., designing a bridge) or purely theoretical (e.g., predicting a solar eclipse or the orbits of heavenly bodies) in nature. One of the fundamental problems here is the design of numerical methods for the determination of the *roots* (or zeroes) of real-valued algebraic functions or, in general, systems of such functions, which themselves may be linear or nonlinear. Since very few of these problems have any known methods that can solve them *directly*, a vast majority of numerical methods are *iterative* in nature; i.e., a current candidate solution is being improved upon until an acceptable answer is found. Iterative techniques, one of the fundamental principles in computer science, may be used to find roots of functions, solutions to systems of linear and nonlinear equations, and solutions to ordinary or partial differential, or integral equations.

In the case of a single linear function, the problem is trivial and can be solved directly. For a system of linear equations, the method of *Gaussian elimination* may be applied to compute the solution. Although not originally named as such, this method first appeared in the chapter on rectangular arrays of a Chinese mathematical book, *Jiuzhang suanshu*, which may have been written as early as 150 BC.

German mathematician Johann Carl Friedrich Gauss (1777–1855), whose contributions to mathematics include linear algebra, number theory, statistics, and many others, came up with this method independently in the early 19th century. Gaussian elimination is an algorithm that may be used for solving systems of linear equations, finding the rank of a matrix, and (if it exists) calculating the inverse of a square matrix.

Whereas Gaussian elimination is a good method that is especially well suited for hand computations, for a vast majority of large-sized or ill-conditioned matrices, more powerful methods are recommended. These include decomposing a given matrix into a product of two triangular matrices, usually called L and U. That is, if A is a nonsingular matrix, then A can be uniquely expressed as a product LU, where L is a lower-triangular matrix with 1's on its main diagonal and U is an upper-triangular matrix.

Then, instead of solving $Ax = b$, we have $LUx = b$, which can now be computed in two steps. First, using *forward* substitution, we obtain a vector y, where $Ly = b$, and next we use *backward* substitution to get x, where $Ux = y$. In both cases, i.e., solving for y and then for x, the computations are performed in $O(n^2)$ steps (See Chapter 26). Of course, the initial decomposition of A into LU itself does require $O(n^3)$ steps, which is also the running time of Gaussian elimination. It is easy to see that if we need to solve $Ax = b$ multiple times for different b's, then the LU decomposition approach is computationally more efficient than a repeated use of Gaussian elimination. Several methods can be used to perform the actual factorization of A; the most popular of these is *Doolittle's* algorithm, but we will not present its details here.

Example 19.1

Solve $Ax = LUx = b$.
   Let

$$A = \begin{pmatrix} 1 & 2 & -1 \\ -2 & -5 & 3 \\ -1 & -3 & 0 \end{pmatrix}, \quad x = \begin{pmatrix} x_1 \\ x_2 \\ x_3 \end{pmatrix}, \quad b = \begin{pmatrix} -1 \\ 5 \\ -2 \end{pmatrix}.$$

Then the equation $Ax = b$ is

$$\begin{pmatrix} 1 & 2 & -1 \\ -2 & -5 & 3 \\ -1 & -3 & 0 \end{pmatrix} \begin{pmatrix} x_1 \\ x_2 \\ x_3 \end{pmatrix} = \begin{pmatrix} -1 \\ 5 \\ -2 \end{pmatrix}$$

and the matrices L and U are

$$L = \begin{pmatrix} 1 & 0 & 0 \\ -2 & 1 & 0 \\ -1 & 1 & 1 \end{pmatrix}, \quad U = \begin{pmatrix} 1 & 2 & -1 \\ 0 & -1 & 1 \\ 0 & 0 & -2 \end{pmatrix}.$$

The solution for y is $y_1 = -1$, $y_2 = 3$, and $y_3 = -6$; the final solution for x is $x_1 = 2$, $x_2 = 0$, and $x_3 = 3$.

For better accuracy of the solution, as well as the computational time required to obtain it in the first place, other decomposition methods (e.g., *QR*), or even iterative approaches, such as the methods of *Jacobi iteration* or *successive overrelaxation*, may also be used.

In case of a nonlinear function of a single variable, e.g., $f(x) = y$, there are a number of iterative methods that may be used to compute the solution. The most basic and intuitive method is that of *bisection*. The bisection method is a root-finding algorithm that repeatedly halves the remaining interval, that is being bracketed by a and b, and then selects the subinterval (in which a root must lie) for further computation. Its convergence is linear, which is relatively slow; however, it is guaranteed to come up with a solution if the initial interval contained it and the function was continuous.

A bracketing method of *false position* or *regula falsi* has all of the advantages of the bisection method; i.e., it is guaranteed to converge, but its convergence to an answer is much quicker. For two bracketing points a and b, instead of finding their midpoint as the next bracket point, this method computes the next point, say c, to be the intersection (with the x-axis) of the line between $f(a)$ and $f(b)$. Depending on the value of $f(c)$, the new point then replaces either a or b as the bracket. The convergence of *regula falsi* is said to be superlinear. The invention of this method has been credited to Indian mathematicians (ca. 200 BC), although it had been mentioned in Chinese books dated from 200 BC. Fibonacci also mentions this method in his 1202 book *Liber Abaci* (The Book of Calculation), in which Europe was introduced to Arabic numerals (and these, themselves, apparently came from India as well; see Chapter 9).

If a function is continuous and its derivative is known, then a quadratically convergent approach, called Newton's (Newton–Raphson) method is most useful for finding its zeroes. Sir Isaac Newton (1643–1727) was an English mathematician and physicist who is widely recognized as one of the most influential scholars in the history of mankind. His many contributions to science include mathematics, including the invention of infinitesimal calculus, which he shares with German mathematician Gottfried Leibniz; physics, in particular, for what are now known as *Newton's three laws of motion*; astronomy; philosophy; and many others. Newton described his method in *De analysi per aequationes numero terminorum infinitas* (1669), but that was not published until 1711. Newton's method, as such, was first described in 1685 in John Wallis's (1616–1703) book *A Treatise of Algebra Both Historical and Practical*.

Newton's method is defined by one of the best-known equations in numerical analysis, and perhaps in all of mathematics [3]:

$$x_{i+1} = x_i - \frac{f(x_i)}{f'(x_i)}.$$

This method often converges remarkably quickly, especially if the iteration begins sufficiently close to the solution, i.e., within what is known as the *radius of attraction*. However, if the initial guess for the solution is far from the actual root, or if at any point during the iteration the derivative is close to zero, and in several other instances, Newton's method can, in fact, diverge, or it may simply oscillate around the final answer. Over the years, many mathematicians have proposed numerous modifications and improvements to Newton's method in order to address these and other problems.

Example 19.2

Below is the numerical computation of $\sqrt{2}$ by Newton's method.
  Since $\sqrt{2}$ is the positive root of $f(x) = x^2 - 2 = 0$, we have $f'(x) = 2x$. Let the initial approximation of the root be $x_0 = 2$. Newton's iterative computation is as follows:

$$x_1 = 2 - f(2)/f'(2) = 2 - (4 - 2)/4 = 1.5$$

$$x_2 = 1.5 - f(1.5)/f'(1.5) = 1.5 - (1.5^2 - 2)/3 \approx 1.4166667$$

$$x_3 = 1.4166667 - f(1.4166667)/f'(1.4166667) \approx 1.4142156$$
$$x_4 = 1.4142156 - f(1.4142156)/f'(1.4142156) \approx 1.4142135.$$

The fundamental idea of the method is as follows: one starts with an initial guess, say x_0, which is hopefully close to the solution. The function then is approximated by its tangent line—hence the need to know its derivative, and the x-intercept point of this tangent line is determined. This point now becomes an improved approximation for the solution. The process continues until a sufficiently good solution is obtained. If the derivative of the function is not known, it may be computationally approximated (at each point of the iteration), in which case this would become the *secant method*.

However, Newton's original description (of what is now known as Newton's method) was not in the iterative form as defined in the above equation. Newton applied his method only to the polynomials, and he did not present it as iterations over successive approximations. He approximated the function itself with polynomials, and only as the final step did he arrive at an approximation for the actual solution, i.e., the root or zero. One might argue that the method (in its original presentation) was, per se, not iterative.

Newton's method was found to have applications as early as in the 17th century. For example, Japanese mathematician Seki Takakazu-Kōwa (1642–1708), also known as *Japan's Newton* [4], used this method to solve equations that came up in his astronomy computations. Since then, countless mathematicians, physicists, engineers, and other scientists have used it to solve problems in their respective areas.

The person who is perhaps best known for modifying Newton's method to use it for finding successively (iteratively) better approximations to the zeroes of functions was an English mathematician, Joseph Raphson (1648–1715). His approach was also an algebraic method, and it was restricted only to polynomials. However, Raphson's method was iterative, in that a sequence of successive approximations to the solution was being constructed, as opposed to Newton's sequence of polynomials, which were approximating only the function itself. In 1740, Thomas Simpson (1710–1761) reformulated Newton's method as an iterative approach for solving general nonlinear equations. It is the formulation by Simpson that is now known as Newton's method.

Arthur Cayley (1821–1895) was first to notice difficulties in generalizing Newton's method to complex roots of polynomials of higher degree. Newton's method can be extended to solve *systems* of (nonlinear) equations. In the 1940s, Leonid Kantorovich (1912–1986) gave the necessary and sufficient conditions for the Newton method to converge to a solution. Kantorovich, a Soviet mathematician, is best known for his work on resource allocation problems, invention of linear programming, and for receiving the 1975 Nobel Prize in Economics for his work. Unfortunately, the process of verifying Kantorovich's conditions appears to be just as hard as the computation of the solution itself; therefore, most Newton-type procedures usually do not check for them.

Newton's method was generalized by Mieczysław Altman (1916–1997), a Polish-American mathematician, in "A Generalization of Newton's Method" [5]. Later, Altman came up with the method of *contractors*, which formed a basis for a unified theory encompassing a large class of iterative numerical methods, including successive approximations, Newton, Newton–Kantorovich, Newton–Altman, as well as steepest descent and other gradient-type methods. This work is described in his book *Contractors and Contractor Directions Theory and Applications: A New Approach to Solving Equations* [6].

19.3 MODERN NUMERICAL ANALYSIS AND ITS PROBLEM DOMAINS

Modern numerical analysis [2] includes the design, convergence rate estimation, error analysis, and stability of the various numerical methods that are being proposed. In general, such analyses are independent of the computer on which they are to be applied. In some instances, however, issues such as word size and floating-point accuracy cannot be ignored, especially if a specific family of methods is being analyzed for a particular class of computers. The ever-increasing size and mathematical complexity of models in physics, engineering, and other domains necessitated the need for a formal approach for analyzing and evaluating the numerical methods that were being programmed to solve these problems.

Although there have been many papers on the subject prior to the 1940s, modern numerical analysis is said to have started in the late 1940s in parallel with the construction of the first programmable electronic computers. John von Neumann's (1903–1957) and Herman Goldstine's (1913–2004) "Numerical Inverting of Matrices of High Order" paper in the *Bulletin of the AMS* (1947) was one of the first to discuss some of the issues in mathematical and error analyses in the context of solving large problems on a computer. Von Neumann, a Hungarian-American mathematician, is credited with many inventions in mathematics and economics, such as *game theory* and the design and architecture of modern-day computers. Today, a vast majority of computers have what is known as *Von Neumann architecture*, in which computers store their programs (and data) in memory and have a program counter pointing to the next instruction about to be executed. Goldstine was an American expert in ballistic computations and was one of the first Electronic Numerical Integrator and Computer (ENIAC) programmers.

Numerical methods are often used to solve various *optimization* problems, where the best operational configurations for models, which are often represented by nonlinear functions, must be computed. Usually, these optimization problems are *constrained* in that the allowable solutions can only come from some bounded domain. Of course, from a computational standpoint, these restrictions can only make the problem more difficult. The calculation of an optimal solution is not, by any means, restricted to finding the zeroes of functions or their derivatives. Many problems, e.g., modeling and the optimization of the flow of messages on the Internet, simply do not have a closed-form functional representation. In such cases, various approximation schemes may be used, including self-adaptive ones, which may adjust the model if the predicted behavior is not in line with the actual one being observed.

Numerical analysis includes other classes of computational methods. For example, given a set of data points, how could one predict the value of a function between any two measurements? This problem is called *interpolation*, and it has numerous applications in sciences, business, and industry. If the number of points is small, one may use a polynomial to represent the predicted value of a given function. Otherwise, piecewise low-degree polynomial or trigonometric functions, called *splines*, are preferable. Numerical methods for the approximation of integrals and derivatives of functions are often based on interpolation. *Simpson's rule* (named after Thomas Simpson; see the previous section) is one of the simpler interpolation-based methods that may be used to compute an area under a curve (i.e., integrate).

The problem of *extrapolation*, or prediction of a value of a function outside the measured range, is, from the mathematical as well as computational standpoint, much more difficult. Even from a purely philosophical point of view, there cannot exist a foolproof method for extrapolation, because if one did exist, we could all use it to predict the future prices in the stock market and everyone would become rich.

The mathematical modeling of physical systems is often accomplished with various forms of (partial) differential and integral equations. Obtaining solutions to these problems is inherently computational and requires well-designed and stable methods. Numerical methods that may be used for solving ordinary or partial differential equations include the *Runge–Kutta, predictor-corrector, finite difference, finite volume,* and *finite element methods*. These, however, are but a few examples of what is a vast number of methods in this area.

The computation of eigenvalues and eigenvectors of matrices, i.e., determination of the real and complex roots of their characteristic polynomial, is yet another area for which numerical methods have been developed. The *power method, Jacobi's, Householder's,* and the *QL method* (latter two for symmetric matrices, which often arise in solving differential equations) may be used for this purpose.

Numerical methods are being applied to address problems in areas as diverse as weather prediction, aircraft design, medical imaging, quantum mechanics, structural engineering, system simulation, and stock market analysis. Today, these methods are implemented on computers whose task it is to perform the necessary calculations.

REFERENCES

1. Numerical Methods, http://www. numericalmathematics.com/numerical_mrthods.html.
2. Wikipedia, Numerical Analysis, http://en.wikipedia.org/wiki/Numerical_methods.
3. Wikipedia, Newton's Method, http://en.wikipedia.org/wiki/Newton%27s_method.
4. Wikipedia,SekiTakakazu,http://en.wikipedia.org/wiki/Seki_Tadaki_Takakazu.
5. M. Altman, A Generalization of Newton's Method, *Bull. Acad. Polon. Sci.*, 3, 1955.
6. M. Altman, *Contractors and Contractor Directions Theory and Applications: A New Approach to Solving Equations*, Marcel Dekker, New York, 1977.

CHAPTER 20

Modular Arithmetic

20.1 CLOCK ARITHMETIC

For addition, subtraction, and multiplication, the result of any of these operations of two integers is also an integer. On the other hand, for a pair of integers, the result of dividing one integer by the other is not necessarily an integer. That is, the set of integers is closed under addition, subtraction, and multiplication, but not under division. However, any of the set of rational numbers, the set of real numbers, and the set of complex numbers is closed under addition, subtraction, and multiplication, as well as division (when dividing by any nonzero element).

For any finite set of integers except for the set consisting of just 0, the set is not closed under any ordinary addition, subtraction, multiplication, and division. People in ancient civilizations devised clock arithmetic on a finite set of integers, which is closed under addition, subtraction, and multiplication. It is an arithmetic system for the set of integers {1, 2, 3, 4, 5, 6, 7, 8, 9, 10, 11, 12}, where numbers wrap around after they reach 12. This kind of arithmetic is called modular arithmetic. The introduction of the first clock began with ancient astronomers noticing the phenomenon of the rising and setting of the sun. The clock arithmetic became known about 5000 years ago when Middle East and North Africa civilizations made the earliest clock to enhance their calendars. Historically, units of time in many civilizations are duodecimal (a positional notation number system of base-12). There are 12 months in a year, and the Babylonians had 12 hours in a day (at some point this was changed to 24 hours) [4].

In the case of clock arithmetic with modulo 12, 12 o'clock is equivalent to 0 o'clock. Clock arithmetic on $\{1, 2, ..., 12\}$ is isomorphic to clock arithmetic on $Z_{12} = \{0, 1, 2, ..., 11\}$. More generally, arithmetic of modulo m can be defined on the set of integers $Z_m = \{0, 1, 2, ..., m-1\}$. For clock arithmetic with modulo m, any integer a is congruent to $a \pm km$ for any integer k. This relation is denoted by $a \equiv_m a + km$, or $a \equiv a + km$ (modulo m). Equivalently, if $a-b$ is evenly divisible by m, we say that a is congruent to b (modulo m). Although in general clock arithmetic is not closed under division, if m is a prime number, then clock arithmetic with modulo m is closed under division (dividing by the nonzero element) as well. In other words, if m is a prime number, then for any nonzero element a in Z_m, the inverse of a (modulo m) exists, and it is denoted by a^{-1} (modulo m) or $1/a$ (modulo m).

Example 20.1

For the 12-hour clock, the day is divided into two 12-hour periods. If the time is 5 o'clock in the morning, then 8 hours later it will be 1 o'clock in the afternoon. Since $5 + 8 = 13 \equiv 1$ (modulo 12), we may say that it will be 13 o'clock.

Suppose that a man drives on a highway. He drove 5 hours and arrived at 2 o'clock in the afternoon. Then he started driving on the highway at 9 o'clock in the morning, since $2-5 = -3 \equiv -3 + 12 = 9$ (modulo 12).

Suppose that a worker starts five jobs sequentially at 4 o'clock in the morning. He spends 3 hours to complete one job. If he works without any rest until he finishes all the jobs, then he will finish all the jobs at 7 o'clock in the evening. Since $4 + 5 \times 3 = 19 \equiv 7$ (modulo 12), we may say that he will finish all the jobs at 19 o'clock or 7 p.m.

Example 20.2

Let Sunday, Monday, Tuesday, Wednesday, Thursday, Friday, and Saturday correspond to 0, 1, 2, 3, 4, 5, and 6, respectively. We can calculate a day of the week by clock arithmetic with modulo 7. Suppose we know that May 2, 2012, is Wednesday. We can calculate that May 25, 2012, is Friday by $3 + (25 - 2) = 26 \equiv 5$ (modulo 7).

Example 20.3

In the addition of modulo 2 arithmetic, $0 + 0 = 0, 0 + 1 = 1, 1 + 0 = 1$, and $1 + 1 = 0$. This addition can be realized by the *exclusive OR* gate. It is equivalent to a logical formula $(\neg x \wedge y) \vee (x \wedge \neg y)$ if true and false correspond to 1 and 0, respectively.

Example 20.4

For clock arithmetic with modulo 5, the inverses of 1, 2, 3, and 4 are 1, 3, 2, and 4, respectively. Notice that $1 \times 1 = 1, 2 \times 3 = 6 \equiv 1, 3 \times 2 = 6 \equiv 1$, and $4 \times 4 = 16 \equiv 1$ (modulo 5). In general, if m is a prime number, then Z_m is also closed under division (dividing by any non-zero element).

As described in Chapter 7, in ancient cryptography, modular arithmetic was used implicitly. For example, according to Caesar cipher, each letter α in a plaintext is encoded as $\alpha + 3$ (modulo 26), and each letter β in a cipher-text is decrypted as $\beta - 3$ (modulo 26), where letters A to Z correspond to 0 to 25, respectively.

20.2 CHINESE REMAINDER THEOREM

Not much is known about the origin of an old Chinese text on mathematics, *Sunzi Suanjing*. It is greatly believed that the book was completed around 400 AD. It consists of three chapters. The first chapter describes measuring systems using counting rods, and methods for calculating multiplication, division, and square roots. The second and third chapters consist of problems (28 and 36, respectively) concerning fractions, areas, volumes, and others [8]. These problems are rather easier than the problems in another ancient text, *Nine Chapters on Mathematical Art* (its origin is believed to be in first century BC or first century AD in China) [9]. However, one problem (problem 26 in Chapter 3) in the *Sunzi Suanjing* is particularly interesting. It is as follows:

> Suppose we have an unknown number of objects. When we continue counting in threes, 2 objects are left over, when we continue counting in fives, 3 objects are left over, and when we continue counting in sevens, 2 objects are left over. How many objects are there?

The problem above is typically modular arithmetic. That is, it can be equivalently described as follows:

There is a number. If it is divided by 3, then the remainder is 2; if it is divided by 5, then the remainder is 3; and if it is divided by 7, then the remainder is 2. What is the number?

A Swiss mathematician, Leonhard Euler (1707–1783), was a pioneer of the modern approach to modular arithmetic. He introduced the idea of being congruent to an integer in 1750. Modular arithmetic was further advanced by Johann Carl Friedrich Gauss (1777–1855) in his book *Disquisitiones Arithmeticae*, published in 1801. The problem in *Sunzi Suanjing* can be described in a modern mathematical form by introducing a congruence relation as the following simultaneous congruence equations:

$$x \equiv 2 \text{ (modulo 3)}$$

$$x \equiv 3 \text{ (modulo 5)}$$

$$x \equiv 2 \text{ (modulo 7)}.$$

The solution to the simultaneous equations above is $x = 23 + 105k$ for any nonnegative integer k, and its smallest solution is 23. Equivalently, we can say that x is a solution to the simultaneous congruence equations above if and only if $x \equiv 23 \text{ (modulo 105)}$. The *Chinese remainder theorem* is a generalization of the problem in *Sunzi Suanjing*. The general solution to the problem was given by Gauss in 1800. A modern statement of the Chinese remainder theorem in algebraic language is as follows:

The Chinese remainder theorem is really a method of solving certain systems of simultaneous congruence equations. Suppose that $m_1, \ldots, m_r$ are pairwise relatively prime positive integers, and that $a_1, \ldots, a_r$ are integers. Consider the following system of simultaneous congruence equations:

$$x \equiv a_1 \text{ (modulo } m_1)$$

$$x \equiv a_2 \text{ (modulo } m_2)$$

$$\vdots$$

$$x \equiv a_r \text{ (modulo } m_r)$$

The Chinese remainder theorem asserts that this system has a unique solution modulo $M = m_1 \times m_2 \times \ldots \times m_r$. For $1 \le i \le r$, define $M_i = M/m_i$. The following theorem describes an efficient algorithm for solving systems of simultaneous congruence equations [1]. Note that M_i and m_i are relatively prime, and the inverse of M_i (modulo m_i) exists ($1 \le i \le r$). The inverse can be efficiently calculated by a variation of Euclidean algorithm (it is called extended Euclidean algorithm [2]).

Theorem 20.1 (Chinese remainder theorem)

Suppose that m_1, ..., m_r are pairwise relatively prime positive integers. Then the system of r congruence equations $x \equiv a_i$ (modulo m_i) ($1 \le i \le r$) has a unique solution modulo $M = m_1 \times \ldots \times m_r$, which is given by

$$x = a_1 M_1 y_1 + a_2 M_2 y_2 + \ldots a_r M_r y_r \text{ (modulo } M)$$

where $M_i = M/m_i$, and $y_i = M_i^{-1}$ (modulo m_i), for $1 \le i \le r$. ▪

Example 20.5

For the problem in the *Sunzi Suanjing* ($x \equiv 2$ (modulo 3), $x \equiv 3$ (modulo 5), $x \equiv 2$ (modulo 7)), $a_1 = 2$, $a_2 = 3$, $a_3 = 2$, $m_1 = 3$, $m_2 = 5$, $m_3 = 7$, $M = 3 \times 5 \times 7 = 105$, $M_1 = 105/3 = 35$, $M_2 = 105/5 = 21$, and $M_3 = 105/7 = 15$. Then $M_1^{-1} \equiv y_1 \equiv 35^{-1} \equiv 2$ (modulo 3), $M_2^{-1} \equiv y_2 \equiv 21^{-1} \equiv 1$ (modulo 5), and $M_3^{-1} \equiv y_3 \equiv 15^{-1} \equiv 1$ (modulo 7). From the Chinese remainder theorem the solution to the problem is $2 \times 35 \times 2 + 3 \times 21 \times 1 + 2 \times 15 \times 1 \equiv 23$ (modulo 105).

Modular arithmetic was also described by Indian mathematicians in the 6th and 7th centuries and a European mathematician in the 13th century. An algorithm for solving the Chinese remainder theorem was given by Aryabhata (476–550) [3, 10]. Special cases of the Chinese remainder theorem were given by Brahmagupta (596–668) [11], and in the book *Liber Abaci* in 1202 [7].

The following problem was originally given by Brahmagupta, known as Brahma's correct system or the egg-woman problem (the solution to this problem is 301):

An old woman went to a market and was selling eggs from her basket. A horse stepped on her basket and crushed the eggs. The rider offered to pay for the damage and asked her, "How many eggs were there in the basket?" She did not remember the exact number of eggs, but when she had taken them out two at a time there was one egg left. The same happened to the remainder when she had taken them out three, four, five, and six at a time, but when she had taken them out seven at a time they came out even. What is the smallest number of eggs she could have had in her basket?

The same problem is also given in Fibonacci's *Liber Abaci*. The following is quoted from the English translation of *Liber Abaci* by L. E. Sigler [7]:

There is a number which when divided by 2, or 3, or 4, or 5, or 6, always has a remainder 1, and it is truly integrally divisible by 7. It is sought what is the number.

Maarten Bullynck conjectures possible routes of how the Chinese remainder problem reached Europe in his paper [6]. The following is quoted from his paper [6]:

In continental Europe, remainder problems show up for the first time in medieval manuscripts on calculation, perhaps through the mediation of Italian merchants returning from China, perhaps through Arabic translations of Indian sources.

20.3 FERMAT'S LITTLE THEOREM

Pierre de Fermat (1601–1665) has been called the greatest amateur mathematician [5]. He communicated mathematical discoveries in numerous letters, usually without proof to his friends. However, he became one of the best mathematicians in his century. Fermat was a pioneer in several areas of mathematics.

One of the theorems discovered by Fermat states that if p is a prime, then for any integer a, $a^p - a$ will be eventually divided by p. This can be expressed in modular arithmetic notation as follows:

$$a^p \equiv a \ (\text{modulo } p).$$

A variant of this theorem is stated in the following form: if p is a prime, and a and p are relatively prime, then the multiplicative order of a is $p - 1$. Hence,

$$a^{p-1} \equiv 1 \text{ (modulo } p).$$

This theorem was stated in a letter to his friend in 1640. He did not prove it, but added the following statement [12]: "This proposition is generally true for all progression and for all primes. I would send you its proof if I were not afraid to be too long."

We call this theorem *Fermat's little theorem*, after an English mathematician, James Joseph Sylvester (1814–1897), who called it by the name to distinguish it from Fermat's last theorem. Euler first published a proof of Fermat's little theorem in 1736, but Gottfried Wilhelm Leibniz (1646–1716) rediscovered and proved the same result in unpublished notes in 1683. The proof by Leibniz is virtually the same as the proof by Euler.

Euler's phi function, $\varphi(n)$, is an arithmetic function that counts the number of positive integers less than n and relatively prime to n. Euler introduced this function in 1760. The standard notation $\varphi(n)$ is from Gauss's paper "Disquisitiones Arithmeticae" in 1801. For a prime number p, $\varphi(p) = p-1$, and for the product of two distinct primes p and q, $\varphi(p \times q) = (p-1)(q-1)$. Let Z_n^* be the set of elements that are relatively prime to n and in Z_n. For example, $Z_6 = \{0, 1, 2, 3, 4, 5\}$ and $Z_6^* = \{1, 5\}$. The following theorem is a variation of Fermat's little theorem. It is more general than the original Fermat's little theorem [12].

Theorem 20.2

If a is in Z_n^*, then $a^{\varphi(n)} \equiv 1$ (modulo n). ■

Fermat's little theorem and its variations play a crucial role in primality testing, and the factorization of polynomials and integers. As we will describe in Chapter 30, these testing algorithms are particularly useful in modern cryptography.

REFERENCES

1. D. E. Knuth, *Seminumerical Algorithms (The Art of Computer Programming* (vol. 2, 2nd ed.), Addison Wesley, Reading, MA, 1981.
2. D. R. Stinson, *Cryptography: Theory and Practice*, CRC Press, New York, 1995.

3. T. R. N. Rao and Chung-Huang Yang, *Modular Arithmetic: From Ancient India to Public-Key Cryptography*, Technical Report, University of Louisiana at Lafayette, 2006.

4. Wikipedia, Modular Arithmetic, http://en.wikipedia.org/wiki/Modular_arithmetic.

5. J. von zur Gathen and J. Gerhard, *Modern Computer Algebra* (2nd ed.), Cambridge University Press, Cambridge, UK, 2003.

6. M. Bullynck, Modular Arithmetic before C. F. Gauss: Systematizations and Discussions on Remainder Problems in 18th Century Germany, *Historia Mathematica*, 36(1), 48–72, 2009.

7. L. E. Sigler, *Fibonacci's Liber Abaci: Leonardo Pisano's Book of Calculation* (English translation of *Liber Abaci*, 1202), Springer, New York, 2002.

8. Lay Yong Lam and Tian Se Ang, *Tracing the Conception of Arithmetic and Algebra in Ancient China*, World Scientific Publication, Singapore, 2004.

9. Wikipedia, The Nine Chapters on the Mathematical Art, http://en.wikipedia.org/wiki/The_Nine_Chapters_on_the_MathematicalArt.

10. Wikipedia, Aryabhata, http://en.wikipedia.org/wiki/Aryabhata.

11. Wikipedia, Brahmagupta, http://en.wikipedia.org/wiki/Brahmagupta.

12. Wikipedia, Fermat's Little Theorem, http://en.wikipedia.org/wiki/Fermat's_little_theorem.

CHAPTER 21

Cybernetics and Information Theory

21.1 NORBERT WIENER AND CYBERNETICS

Norbert Wiener (1894–1964) studied mathematics at Tufts College from 1906 to 1909, zoology at Harvard University from 1909 to 1910, and philosophy at Cornell University from 1910 to 1911. Then, he returned to Harvard University, while continuing his philosophy studies. Wiener had been interested in the scientific method for a long time. He was a participant in a Harvard seminar run by Josia Royce (1855–1916) between 1911 and 1913. Harvard University awarded Wiener a Ph.D. in 1912, when he was 18 years old, for his dissertation on mathematical logic [10].

In 1914, Wiener traveled to Europe, where he studied under the guidance of Bertrand Russell (1872–1979) and G. H. Hardy (1877–1947) at Cambridge University, and David Hilbert (1862–1943) and Edmund Landau (1877–1938) at the University of Göttingen. During 1915–1916, he taught philosophy at Harvard University, and then worked as an engineer for General Electric Co. In 1926, Wiener returned to Europe, and spent most of his time at Göttingen and at Cambridge, where he worked on Brownian motion, Fourier integrals, the Dirichlet problem, harmonic analysis, and Tauberian theorems.

In the 1940s, during and after World War II, Wiener and Arturo Rosenblueth (1900–1970) of the Harvard Medical School conducted a monthly series of discussion meetings on the scientific method. The participants were mostly young scientists from Harvard Medical School.

Wiener played a key role in the entire field of the theory of control and communication in machines and animals in this series of meetings. Rosenblueth's return to Mexico in 1944 and the general confusion after World War II ended this series of meetings [1].

During World War II, Wiener also worked on guided missile technology, and studied how sophisticated electronics used the feedback principle. He noticed that feedback is a key feature of life-forms, from the simplest plants to the most complex animals, changing their actions in response to stimuli from their environment. Wiener developed this concept into the field of *cybernetics*, concerning the combination of animals and machines. In a related work, Wiener investigated information theory independently of Claude E. Shannon (1916–2001), and invented what is now known as the *Wiener filter*. The Wiener filter is a statistically designed filter to reduce the amount of noise present in a signal. The equivalent filter was also derived independently in 1941 by Russian mathematician Andrey Nikolaevich Kolmogorov (1903–1987). Their theory is often called the Wiener–Kolmogorov filter theory.

In 1948, Wiener published a book representing the outcome, after more than a decade, of research undertaken jointly with Rosenblueth at Harvard Medical School. He coined the word *cybernetics* with its current meaning, as defined in the title of his book *Cybernetics or Control and Communication in the Animal and the Machine* [1, 10]. The term *cybernetics* stems from the Greek word *kybernetes* that means "steersman, governor, pilot, or rudder." The book *Cybernetics* discusses Newtonian and Bergsonian time; groups and statistic machines; time series, information, and communication; feedback and oscillation; computer machines and nervous systems; gestalt and universals; cybernetics and psychopathology; and information, languages, and society. In the second edition of the book, published in 1961, Wiener added supplemental chapters on learning and self-reproducing machines and on brain waves and self-organizing systems.

Cybernetics is a broad field of science and technology. The essential goal of cybernetics is to understand and define the functions and processes of systems with feedback loops. British scientist Stafford Beer (1926–2002) called cybernetics the science of effective organization, and another British scientist, Gordon Pask (1928–1996), extended it to include information flow in all media, from stars to brains. It includes the study of feedback, black boxes, and derives concepts such as communication and control in living organisms and machines. French mathematician and computer scientist Louis Couffignal (1902–1966) characterized cybernetics as the art of ensuring the efficiency of actions. American mathematician and computer scientist Louis Kauffman (1945–) proposed that "cybernetics is the study of systems and processes that interact with themselves and produce themselves from themselves." [9] Concepts studied by cyberneticists also include learning, cognition, and adaptive theory. Other fields influenced by cybernetics include, but are not limited to, game theory, system theory, psychology, neurology, brain science, and anthropology.

21.2 SHANNON'S INFORMATION THEORY

Claude E. Shannon (1916–2001) is the celebrated Father of Information Theory. He graduated from MIT with an M.S. in electrical engineering in 1937 and a Ph.D. in mathematics in 1940. He later became a research fellow at the Institute for Advanced Study at Princeton University and joined Bell Laboratories in 1941 [2, 7, 8].

During World War II, Shannon was employed by Bell Laboratories, working on top-secret defense projects in cryptography. His team worked on antiaircraft devices that observe enemy planes or missiles and calculate the trajectories of intercepting missiles. In an early paper, Shannon acknowledged the profound influence he received from Harry Nyquist (1889–1976) and R. V. L. Hartley (1888–1970), who were pioneers and fundamental contributors to data transmissions. In 1943, Shannon met British cryptanalyst and mathematician Alan M. Turing. Shannon was interested in speech encryption, and developed its binary encoding system. In 1945, it occurred to Shannon that the problem of smoothing the data in firing control could be formally treated with some analogy to the problem of separating a signal from interfering noise in communication systems. His work during World War II is closely related to his later publications on communication theory.

In 1948, Shannon's brilliant memorandum appeared as "A Mathematical Theory of Communication" in two parts in the issues of the *Bell System Technical Journal* in 1948 and 1949 [3, 4]. *The Mathematical Theory of Communication*, a book coauthored with Warren Weaver (1894–1978) [5], was published in 1949 as reprints of Shannon's articles and Weaver's memorandum on communication theory. Shannon's paper focuses on the problem of how best to encode the information that a sender wants to transmit [5]. In his revolutionary paper, Shannon introduced a quantitative model of communication as a statistical process underlying information theory. If a message specifies one from a set of n possible messages, according to Shannon's theory, the quantity of the message is defined as $\log_2 n$ binary digits (briefly *bits*). Equivalently, if the probability that the message specifies a fact is $p = 1/n$, then the information of the message is $-\log_2 p = \log_2 n$ bits. The word *bit* was first introduced by John W. Tukey (1915–2000) while he worked with John von Neumann on early computer designs. As Shannon described in his article, the logarithmic measure on messages is the most natural choice for the following reasons:

1. Parameters of engineering importance, such as time, bandwidth, number of relays, etc., tend to vary linearly with the logarithm of the number of possibilities.

2. It is natural to our intuitive feeling as the proper measure. For example, two punched cards should have twice the information capacity of one card, and two identical channels should have twice the information transmitting capacity.

3. Many of the limiting operations are simple in terms of the logarithm but would require clumsy restatement in terms of the number of possibilities.

Shannon showed how information could be quantified with absolute precision. Telephone signals, texts, radio waves, and television pictures could be encoded in bits. The channel capacity of a communication line could be precisely measured in bits. In his fundamental work, he used tools in probability theory, developed by Norbert Wiener and others, and applied them to communication theory. The revolutionary idea by Shannon was promptly adopted by communication and computer engineers. His theory has been widely used to measure computer storage in bits, needed for pictures, voices, and other types of data.

In his article, Shannon represented a discrete information source as a stochastic process. He defined a quantity that would measure how much information is produced by such a process and at what rate this information is produced. Suppose a set of possible events whose probabilities of occurrence are $p_1, p_2, \cdots, p_n$. In the case where these probabilities are known, Shannon introduced a measure called *entropy*:

$$H(p_1, p_2, \cdots, p_n),$$

indicating how much certainty is involved in the selection of the event or how uncertain we are of the outcome. The entropy of the system is defined to be as follows (See Figure 21.1):

$$H(p_1, p_2, \cdots, p_n) = -(p_1 \log_2 p_1 + p_2 \log_2 p_2 + \cdots + p_n \log_2 p_n).$$

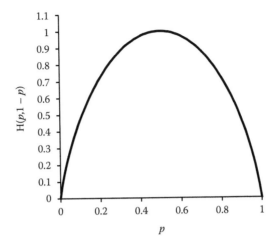

FIGURE 21.1 The entropy of a single coin toss with probabilities $(p, 1-p)$.

As Shannon describes in his paper, it is reasonable to require $H(p_1, p_2, \cdots, p_n)$ possessing the following properties:

1. H should be continuous in p_i (for each i).

2. If all p_i are equal, $p_i = 1/n$, then H should be a monotonic function of n. With equally likely events there is more choice, or uncertainty, when there are more possible events.

3. If a choice is broken down into two successive choices, the original H should be the weighted sum of the individual values of H.

According to the definition of *Shannon entropy*, a single toss of a fair coin has entropy of 1 bit. Two tosses have entropy of 2 bits. The entropy rate for the coin is 1 bit per toss. However, if the coin is not fair, then the uncertainty is lower, and thus Shannon entropy is lower. A series of tosses of a two-headed coin will have zero entropy since the outcomes are entirely predictable.

The word *entropy* in information theory came from the close resemblance between Shannon's formula and the known formula for thermodynamics [6]. In statistical thermodynamics, the most general formula for thermodynamics entropy is given as follows:

$$-K_B(p_1 \log_e p_1 + p_2 \log_e p_2 + \cdots + p_i \log_e p_i + \cdots).$$

where K_B is the Boltzmann constant, and each p_i is the probability of a microstate. This formula was given by Josiah Willard Gibbs (1839–1903) in 1878 and is called *Gibbs entropy*.

21.3 SHANNON–FANO CODING AND HUFFMAN CODING

An efficient coding technique of messages was proposed in Shannon's paper in 1948. Robert Mario Fano (1917–) developed Shannon's method and published it as a technical report [11]. It is a technique for constructing a code based on a set of symbols and their estimated probabilities, and was named *Shannon–Fano coding*. Here, the set of symbols is divided into two sets whose total probabilities are as close as possible to being equal. Then all symbols in the first set are assigned a 0 and all symbols in the second set are assigned a 1. As long as any sets with more than one member remain, the same process is repeated. When a set has been reduced to one symbol, the code generating process is complete. In this way, the code of each symbol is determined as successive binary digits. Note that the code of any symbol does not form the prefix of the code of any other symbol. Shannon–Fano coding can be more formally described via the following encoding algorithm:

Algorithm: Shannon–Fano Coding

```
create a table providing frequencies of symbols;
sort symbols according to frequency in descending order;
start with the entire table;
    division:
        seek pointer to the first and last symbols of
            the segment;
        divide the segment into two parts, both nearly
            equal in sum of frequencies;
        concatenate a binary 0 to the end of each code
            word of the upper part and a binary 1 to the
            end of the code word of the lower part;
        search for the next segment containing at least
            two symbols and repeat division;
coding of the symbols according to the code words
    created in the table;
```

In general, Shannon–Fano coding does not achieve the lowest possible code word length, but it guarantees that all code word lengths are within

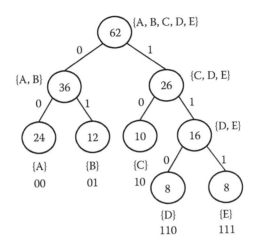

FIGURE 21.2 An encoding by Shannon–Fano coding.

1 bit of their theoretical ideal. The next example shows the construction of Shannon–Fano coding for a small alphabet {A, B, C, D, E} as shown in Figure 21.2.

Example 21.1

Suppose that the frequencies of each symbol of {A, B, C, D, E} are as given in the following table:

Symbol	A	B	C	D	E
Frequency	24	12	10	8	8

The set of symbols {A, B, C, D, E} is divided into two segments, {A, B} and {C, D, E}. Then segment {A, B} is divided into {A} and {B}, and segment {C, D, E} is divided into {C} and {D, E}. Then segment {D, E} is divided into {D} and {E}. The code word of each symbol is given in the following table:

Symbol	A	B	C	D	E
Code word	00	01	10	110	111

The average bit number per symbol is

$$(2 \times 24 + 2 \times 12 + 2 \times 10 + 3 \times 8 + 3 \times 8)/62 \approx 2.258 \text{ (bits per symbol)}.$$

In 1951, Professor Robert M. Fano assigned a term paper on the problem of finding the most efficient binary coding in the information theory

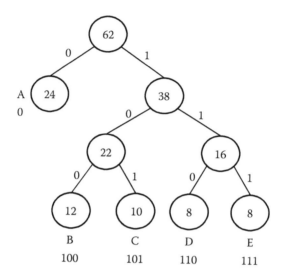

FIGURE 21.3   An encoding by Huffman coding.

course at MIT. David Albert Huffman (1925–1999), a Ph.D. student at that time, started studying the problem, and eventually got the idea of using a frequency-sorted binary tree construction. Huffman avoided the major flaw of the suboptimal Shannon–Fano coding by constructing the tree from the bottom up instead of from the top down. He proved the coding method to be the minimum redundancy binary coding [12] (Figure 21.3). The following algorithm constructs the Huffman coding:

Algorithm: Huffman Coding

```
create table providing frequencies of symbols;
sort symbols according to frequency in descending order;
repeat
    search for the two nodes providing the lowest
        frequencies, which have not been assigned
        a parent node, and assign a parent node with
        a frequency that is the sum of the two
        lower elements
until all nodes are combined together in a root node;
{to generate a Huffman code word we traverse the
constructed tree from the root node to a leaf node,
outputting a 0 every time we take a left hand branch,
and a 1 every time we take a right hand branch}
```

Example 21.2

We use the same frequency table as in Example 21.1. The two lowest-frequency symbols are E and D. These two nodes are connected first, and their parent is created with frequency 16 (sum of frequency 8 of E and frequency 8 of D). Next, we choose C and B and connect them. Their parent with frequency 22 (sum of 10 and 12) is created. Then the parent of E and D, and the parent of C and D are connected, and their parent with frequency 38 (sum of 16 and 22) is created. Leaf node A with frequency 24, and the parent with frequency 38 (the ancestor of B, C, D, E) are connected, and their parent with frequency 62 is created. According to the constructed tree, we assign code words to the set of symbols as shown in the following table:

Symbol	Frequency	Code Word	Code Length	Total Length
A	24	0	1	24
B	12	100	3	36
C	10	101	3	30
D	8	110	3	24
E	8	111	3	24

The average bit number per symbol is

$$(24 + 36 + 30 + 24 + 24)/62 \approx 2.2258.$$

Both Shannon–Fano coding and Huffman coding are prefix codes (sometimes called prefix-free codes). That is, the bit string representing any symbol is never a prefix of the bit string representing any other symbol. Shannon–Fano coding does not offer the best code efficiency. It provides a result similar to that of Huffman coding, but it will never exceed Huffman coding.

21.4 MORSE CODE

One important feature of Shannon–Fano coding and Huffman coding is that the length of the code word for a symbol is approximately inversely proportional to its frequency. More than 100 years before Shannon–Fano coding appeared, this feature was already used in the Morse code. Samuel F. B. Morse (1791–1872) was an American contributor to the invention of a single-wire telegraph system and an inventor of Morse code. Morse was born in Charlestown, Massachusetts. He went to Yale College, where he studied religious philosophy, mathematics, and science. In 1810, he graduated from Yale College and became a professional painter. In the 1930s Morse became interested in electromagnetism and developed the concept of a single-wire telegraph. In 1844, the telegraph wire line was officially opened from Baltimore to the Capitol Building in Washington, D.C. On May 24, 1844, Morse, in the U.S. Supreme Court Chambers in Washington, D.C., sent by telegraph to his colleague Alfred Vail (1807–1859) the famous words, "What hath God wrought" (What has God worked?). In 1845 Morse, his colleagues, and a small group of investors formed the Magnetic Telegraph Company. The first commercialized telegraph line was completed between Washington, D.C., and New York City in the spring of 1846 [13, 14].

Alfred Vail played an important role in the invention of the Morse code. A related code for Morse's telegraph was originally created by Vail in the early 1840s. This code was the forerunner of Morse code. In the 1890s it began to be extensively used for early radio communication before it became possible to transmit voice messages. In the late 19th and the early 20th century, most high-speed communication systems used Morse code on telegraph lines through undersea cables or by electromagnetic waves.

Morse code is a method of transmitting textual information as a series of on-off tones that can be understood by a skilled listener. International Morse code encodes the Roman alphabet and the Arabic numerals as standardized sequences of short and long signals called dots and dashes. Each character (letter or numeral) is represented by a unique sequence of dots and dashes. The duration of a dash is three times the duration of a dot. Each dot or dash is followed by a short silence [13]. Table 21.1 summarizes the international Morse code:

REFERENCES

1. N. Wiener, *Cybernetics or Control and Communication in the Animal and the Machine*, Hermann & Cie, Paris, 1948.
2. C. E. Shannon, A Symbolic Analysis of Relay and Switching Circuits, *Transactions of the American Institute of Electrical Engineering*, 57, 713–723, 1938.
3. C. E. Shannon, A Mathematical Theory of Communication, *Bell System Technical Journal*, 27, 379–423, 1948.
4. C. E. Shannon, A Mathematical Theory of Communication (Part II), *Bell System Technical Journal*, 28, 623–656, 1949.
5. C. E. Shannon and W. Weaver, *The Mathematical Theory of Communication*, University of Illinois Press, Champaign, 1949.
6. H. Rheingold, *Tools for Thought*, MIT Press, Cambridge, MA, 2000.
7. Claude Shannon, http://www.sprangle.com/steve/shannon.html.
8. Wikipedia, Shannon, http://en.wikipedia.org/wiki/Claude_Shannon.

TABLE 21.1   International Morse Code

A:	· —	U:	· · —
B:	— · · ·	V:	· · · —
C:	— · — ·	W:	· — —
D:	— · ·	X:	— · · —
E:	·	Y:	— · — —
F:	· · — ·	Z:	— — · ·
G:	— — ·		
H:	· · · ·	1:	· — — — —
I:	· ·	2:	· · — — —
J:	· — — —	3:	· · · — —
K:	— · —	4:	· · · · —
L:	· — · ·	5:	· · · · ·
M:	— —	6:	— · · · ·
N:	— ·	7:	— — · · ·
O:	— — —	8:	— — — · ·
P:	· — — ·	9:	— — — — ·
Q:	— — · —	0:	— — — — —
R:	· — ·		
S:	· · ·		
T:	—		

9. Wikipedia, Cybernetics, http://en.wikipedia.org/wiki/Cybernetics.
10. Wikipedia, Norbert Wiener, http://en.wikipedia.org/wiki/Cybernetics.
11. R. M. Fano, *Transmission of Information*, MIT Press, Cambridge, MA, 1961.
12. D. A. Huffman, A Method for the Construction of Minimum-Redundancy Codes, *Proceedings of IRE*, 40, 1098–1101, 1952.
13. Wikipedia, Morse Code, http://en.wikipedia.org/wiki/Morse_Code.
14. Wikipedia, Samuel Morse, http://en.wikipedia.org/wiki/Samuel_Morse.

CHAPTER 22

Error Detecting and Correcting Codes

22.1 PARITY CHECK CODES

We can enjoy chats with friends even in noisy pubs. Even if we cannot hear some sounds of words, we can usually understand each other. Since any language has some redundancy, we can usually guess the meaning of sentences containing some missing sounds or missing words if the missing sounds/words are not too many, or we can ask, "Pardon me, could you please repeat what you said?" These kinds of techniques are also useful in data communications and data processing in digital circuits.

A parity bit in a binary code word is a bit added to ensure that the number of bits with the value 1 in the code word is even or odd. Parity bits are used as the simplest form of error detecting codes. If the code consists of code words with an even (odd) number of 1's, it is called an even (odd) parity check code. In the case of an even (odd) parity check code, given k bits of information, an extra bit is added so that the total number of 1's in a code word is even (odd). For example, in an even parity check code with 4 information bits, the addition of the parity check bit at the beginning consists of the following 16 code words:

0: 00000	4: 10100
1: 10001	5: 00101
2: 10010	6: 00110
3: 00011	7: 10111

8: 11000   12: 01100
9: 01001   13: 11101
10: 01010   14: 11110
11: 11011   15: 01111

Parity check codes were used on magnetic data storage, punched tape data in data communication and processing systems in the early 1950s. For example, on the systems sold by the British company ICL in the 1950s, the 1-inch-wide paper tape had eight hole positions running across it, with the eighth being reserved for a parity bit.

In the case of an even parity check code, if an odd number of bits is changed in transmission, the message will change parity and the error can be detected. The most common interpretation is that a parity value of 1 indicates the existence of an odd number of errors in the data, and a parity value of 0 indicates no errors or an even number of errors in the data.

Parity checking is not robust, since if the number of bits changed is even, the error will not be detected. Moreover, parity does not indicate which bit contains an error, even when it can detect the existence of an error. The data must be discarded entirely and retransmitted from the sender, whenever an error is detected. Consequently, on a noisy transmission medium, a successful transmission could take a long time. However, while the quality of parity checking is poor, this method results in the least overhead.

22.2 HAMMING CODES

The history of error correcting codes began with the publication of famous papers by C. E. Shannon (1948, 1949) [14, 15]. Shannon's information theory told us about the existence of error correcting codes, but he did not tell us how to find such codes. Throughout the 1950s much effort was devoted to finding explicit constructions for classes of codes that would produce an arbitrarily small probability of error, as promised by Shannon.

The first big progress toward the construction of such codes involved *block codes*, which have a strong algebraic flavor. In a block code, each message block with m digits is encoded into a longer sequence of n digits (n-tuple) for fixed m and n ($m < n$). Only certain selected n-tuples, called *code blocks* or, more commonly, *code words*, are transmitted from the sender. At the receiver side, a decision concerning the code word transmitted from the sender is made by the nature of a best guess on the basis of available information. With a good code, the probability of a wrong decision may be much smaller than the probability that the original code word is reproduced without an error at the receiver side. One of the earliest error correcting block codes was introduced in 1950, when Richard Wesley Hamming (1915–1998) described a class of single error correcting codes [5].

Hamming was an American mathematician. He received his bachelor's degree from the University of Chicago in 1937, a master's degree from the University of Nebraska in 1939, and a Ph.D. from the University of Illinois at Urbana-Champaign in 1942 [16]. From 1946 to 1976, he worked at Bell Laboratories, where he collaborated with Claude E. Shannon.

In his paper, Hamming explained that he was led to this topic from a consideration that a large number of operations must be prepared without a single error in the end result. He also described that in a digital computer, a single failure usually means a complete failure, in the sense that if it is detected, no more computing can be performed until the failure is located and corrected. Hamming introduced a distance, or as it is usually called, a metric, in the vector space of 2^n points (i.e., 2^n binary sequences of length n, hereafter denoted by Q^n). The metric is now known as the *Hamming distance*, where the distance between two words is defined to be the number of positions in which the words differ. The definition of the Hamming distance is based on the observation that a single error in a code word changes one coordinate (position), two errors, two coordinates, and in general k errors produce differences in k coordinates. More formally, the Hamming distance between x in Q^n and y in Q^n, $d(x, y)$ is defined by

$$\left|\left\{ i \mid 1 \le i \le n, \ x_i \ne y_i \right\}\right|,$$

where $x = x_1 \cdots x_n$, $y = y_1 \cdots y_n$, and $|S|$ is the number of elements in the set S.

A block code of length n is a subset of Q^n. Let C be a block code. The minimum distance of a code C is the smallest Hamming distance between distinct code words in C, and it is important in determining the error correcting capability of C. It is formally defined as

$$min\{d(x, y) \mid x \text{ in } C, y \text{ in } C, \text{ and } x \ne y\}.$$

Example 22.1

Suppose that we encode 2-bit messages into 5-bit code words as follows:

$$00 \Leftrightarrow 10101, 01 \Leftrightarrow 10010, 10 \Leftrightarrow 01110, 11 \Leftrightarrow 11111.$$

This code is a block code of length 5. The code consists of four code words, 10101, 10010, 01110, and 11111. For this code, the Hamming distances between the specific pairs of distinct code words are:

$$d(10101, 10010) = 3, \quad d(10101, 01110) = 4$$

$$d(10101, 11111) = 2, \quad d(10010, 01110) = 3$$

$$d(10010, 11111) = 3, \quad d(01110, 11111) = 2.$$

The minimum distance of this code is 2. This code has an error detecting capability; however, it does not have any error correcting capability. For example, if the receiver receives 10111, then the receiver cannot specify the error location, although he or she can detect that the received word 10111 contains an error. The original message might have been either 10101 or 11111, with the probabilities of each one taking place being equal. This example shows that a code is incapable of correcting up to t errors unless the minimum distance of the code is at least $2t + 1$.

A *systematic block code* of length n is defined as a block code in which each code word has exactly n binary digits, where m digits are associated with the information, while the other $k = n-m$ digits are used for error detection and correction. This produces a redundancy R, which is defined as the ratio of the number of binary digits used to the absolute minimum number necessary to convey the same information (i.e., $R = n/m$).

Hamming constructed systematic block codes that can correct any single-bit error. These codes are called the *Hamming codes*. For each k ($k \geq 2$) there exists a Hamming code of code word length $n = 2^k-1$, where k is the number of parity check bits and $m = n-k$ is the number of information bits in each code word. The minimum distance of each Hamming code is 3. Let us consider $Q^n = \{0, 1\}^n$, the set of a binary sequence of length n. A block code of length n is a subset of Q^n. Now, visualize a sphere about each of the code words of a code, each sphere with the same radius. Allow these spheres to increase in radius by an integer amount until they cannot be made larger without causing some spheres to intersect. The value of the radius is equal to the number of errors that can be corrected by the code. A perfect code of block length n is one for which there are equal-radii spheres about the code words that are disjoint and that completely fill the space $Q^n = \{0, 1\}^n$. It is a well-known fact that any Hamming code is a perfect code [7, 8, 12].

Example 22.2

There are $2^7 = 128$ binary sequences of length 7. Let us consider the following set of code words:

0000000	0001011	0010110	0011101
0100111	0101100	0110001	0111010
1000101	1001110	1010011	1011000
1100010	1101001	1110100	1111111

This set is a Hamming code of block length $2^3 - 1 = 7$. The first 4 bits and the last 3 bits of each code word are information bits and parity check bits, respectively. That is, $m = 4$, $k = 3$, and $n = 2^3 - 1 = 7$. For the information bits i_1, i_2, i_3, and i_4, the parity check bits p_1, p_2, and p_3 are determined by the following equations:

$$p_1 \equiv i_1 + i_2 + i_3 \text{ (modulo 2)}$$

$$p_2 \equiv i_2 + i_3 + i_4 \text{ (modulo 2)}$$

$$p_3 \equiv i_1 + i_3 + i_4 \text{ (modulo 2)}.$$

For this Hamming code, we can encode any 4-bit message into a 7-bit code word according to the equations above. The minimum distance of this code is 3. The number of binary sequences within the sphere of radius 1 about each code word is $1 + 7 = 8$. Notice that the number of 7-bit sequences of Hamming distance 1 from each code word is 7. These seven binary sequences and the code word itself are the binary sequences within the sphere of radius 1 about the code word. These spheres are disjoint, and their union completely covers the whole space of binary sequences of length 7, since $8 \times 16 = 128$. Therefore, we can conclude that this code is a perfect code. For a received word $a_1 a_2 a_3 a_4 b_1 b_2 b_3$, the following (s_1, s_2, s_3) is called its syndrome:

$$s_1 \equiv a_1 + a_2 + a_3 + b_1 \text{ (modulo 2)}$$

$$s_2 \equiv a_2 + a_3 + a_4 + b_2 \text{ (modulo 2)}$$

$$s_3 \equiv a_1 + a_3 + a_4 + b_3 \text{ (modulo 2)}.$$

For this Hamming code we can decode any received word by the following set of rules:

1. There are no errors in the received word if $s_1s_2s_3 = 000$.
2. b_1 is incorrect if $s_1s_2s_3 = 100$.
3. b_2 is incorrect if $s_1s_2s_3 = 010$.
4. b_3 is incorrect if $s_1s_2s_3 = 001$.
5. a_1 is incorrect if $s_1s_2s_3 = 101$.
6. a_2 is incorrect if $s_1s_2s_3 = 110$.
7. a_3 is incorrect if $s_1s_2s_3 = 111$.
8. a_4 is incorrect if $s_1s_2s_3 = 011$.

22.3 LINEAR CODES

Under component-wise vector addition and component-wise scalar multiplication, the set of n-tuples of elements from $GF(q)$ is a vector space called $GF(q)^n$, where $GF(q)$ is the finite field with q elements (also called the *Galois field* with q elements). A *linear code* is a subspace of $GF(q)^n$. Most of the known good codes belong to a class of codes called linear codes. For ease of explanation, we restrict our attention mainly to the linear block codes over a vector space $GF(2)^n$. Any pair of vectors of the vector space can be added by modulo 2 addition in each component. The modulo 2 addition in $GF(2)$ is defined as:

$$0 + 0 = 1 + 1 = 0 \text{ and } 0 + 1 = 1 + 0 = 1,$$

and the product in $GF(2)$ is defined as

$$1 \times 1 = 1 \text{ and } 0 \times 0 = 0 \times 1 = 1 \times 0 = 0 \times 0 = 0.$$

A linear code is a subspace of $GF(2)^n$. That is, a linear code of length n is a nonempty set of n-tuples over $GF(q)$, called code words, such that the sum of any pair of code words is also a code word, and the product of any code word by an element of $GF(q)$ is also a code word. For any linear code, the all-zero word is always a code word, and any linear combination of code words is also a code word.

The theory of vector spaces can be used to study linear codes. Any set of basis vectors for the subspace can be expressed as rows forming an m by n matrix G called the *generator* matrix of the linear code of length n with m information digits. The set of basis vectors is linearly independent, and the row space of G is a linear code. The set of q^m code words generated by the G is called an (n, m) *linear code* [2, 3, 7, 8]. Given an $m \times n$ matrix A, the $n \times m$ matrix obtained by interchanging the rows and columns of A is called the transpose of the matrix A and is denoted by A^T. Any Hamming code is a linear code. As shown in the following example, the (7, 4) Hamming code given in Example 22.2 is a linear binary code of length 7 with 4 information bits and 3 parity check bits.

Example 22.3

The set of 0001011, 0010110, 0100111, and 1000101 is a set of basis vectors of the (7, 4) binary Hamming code, given in Example 22.2. The 4 by 7 matrix consisting of these row vectors is a generator matrix of the (7, 4) Hamming code. In fact, each code word of the (7, 4) Hamming code is a linear combination of these row vectors as shown below:

$$0000000 = 0 \times (0001011) + 0 \times (0010110) + 0 \times (0100111) + 0 \times (1000101)$$
$$0001011 = 1 \times (0001011) + 0 \times (0010110) + 0 \times (0100111) + 0 \times (1000101)$$
$$0010110 = 0 \times (0001011) + 1 \times (0010110) + 0 \times (0100111) + 0 \times (1000101)$$
$$0011101 = 1 \times (0001011) + 1 \times (0010110) + 0 \times (0100111) + 0 \times (1000101)$$
$$0100111 = 0 \times (0001011) + 0 \times (0010110) + 1 \times (0100111) + 0 \times (1000101)$$
$$0101100 = 1 \times (0001011) + 0 \times (0010110) + 1 \times (0100111) + 0 \times (1000101)$$
$$0110001 = 0 \times (0001011) + 1 \times (0010110) + 1 \times (0100111) + 0 \times (1000101)$$
$$0111010 = 1 \times (0001011) + 1 \times (0010110) + 1 \times (0100111) + 0 \times (1000101)$$
$$1000101 = 0 \times (0001011) + 0 \times (0010110) + 0 \times (0100111) + 1 \times (1000101)$$
$$1001110 = 1 \times (0001011) + 0 \times (0010110) + 0 \times (0100111) + 1 \times (1000101)$$
$$1010011 = 0 \times (0001011) + 1 \times (0010110) + 0 \times (0100111) + 1 \times (1000101)$$
$$1011000 = 1 \times (0001011) + 1 \times (0010110) + 0 \times (0100111) + 1 \times (1000101)$$
$$1100010 = 0 \times (0001011) + 0 \times (0010110) + 1 \times (0100111) + 1 \times (1000101)$$
$$1101001 = 1 \times (0001011) + 0 \times (0010110) + 1 \times (0100111) + 1 \times (1000101)$$
$$1110100 = 0 \times (0001011) + 1 \times (0010110) + 1 \times (0100111) + 1 \times (1000101)$$
$$1111111 = 1 \times (0001011) + 1 \times (0010110) + 1 \times (0100111) + 1 \times (1000101)$$

In a linear code, one-to-one correspondence of m-tuples (i.e., a sequence of m information digits) and code words can be used as an encoding procedure, but the most natural way is to use the following transformation by the product of vector i (an m-tuple of information digits) and an m by n generator matrix G:

$$c = iG,$$

where c is the code word corresponding to i.

Because a linear code is a subspace, it has an orthogonal complement, which is the set of all vectors orthogonal to the set of code words. The orthogonal complement is also a subspace. The orthogonal complement has a dimension of $n-m$, and its basis has $n-m$ vectors. Let H be a matrix with these basis vectors as rows. Then an n-tuple c is a code word of the linear code if and only if it is orthogonal to every row vector of H. That is,

$$cH^T = 0,$$

where H^T is the transpose of H.

This gives us a way for testing whether a received word is a code word. The $(n-m)$ by n matrix H is called a *parity check matrix* of the code.

We next explain the Golay code, which is also a perfect code. Notice that

$$\left({}_{23}C_0 + {}_{23}C_1 + {}_{23}C_2 + {}_{23}C_3 \right) \times 2^{12} = 2^{23},$$

where ${}_nC_r$ denotes the number of combinations of n objects taken r at a time. For example, the combinations of the letters a, b, c, and d taken three at a time are

$$\{a, b, c\}, \{a, b, d\}, \{a, c, d\}, \{b, c, d\}.$$

Thus ${}_4C_3 = 4$. The equation above is a necessary (but not sufficient) condition for the existence of a perfect linear code (23, 12) that can correct up to triple errors over $GF(2)$. Marcel J. E. Golay (1902–1989) found such a code in 1949 [4]. In the binary Golay code, there are 2^{12} code words, and the number of binary sequences of length 23 within the radius 3 sphere at each code word as a center is equal to the sum of the numbers in the brackets of the equation above. These spheres do not overlap, and the union of them completely covers the vector space of 2^{23} binary sequences. That is, the binary Golay code is a triple error correcting linear code and a perfect code as well. Notice that the minimum distance of the binary Golay code is 7.

As described in the previous section and this section, the study of linear codes began with the early papers of Hamming (1950) and Golay (1949) [4, 5]. Most of the algebraic setting of linear codes is from the 1956 paper [10] by David S. Slepian (1923–2007). Earlier, Zen-ichi Kiyasu (1915–2006) had noticed the relationship between linear codes and subspaces of vector spaces in 1953 [6]. A. Tietavainen and J. H. van Lint proved that there exist no linear (nontrivial) perfect codes other than the Hamming codes and the Golay code, in 1974 and in 1975, respectively [11, 12].

Reed–Muller codes are also a class of linear codes over $GF(2)$ that are easy to describe and can be decoded by a simple voting technique. The Reed–Muller codes were discovered by David E. Muller (1924–2008) in 1954 [17], and in the same year Irving S. Reed (1923–2012) discovered the decoding algorithm for them [18].

The class of cyclic codes is a subclass of the class of linear codes obtained by imposing an additional structural requirement on the codes. An (n, m) linear code C is called a *cyclic code* if it has a property that any cyclic shift of a code word of C is also a code word of C. More formally, if an n-tuple

$$\mathbf{v} = v_0 v_1 v_2 \cdots v_{n-1}$$

is a code word of C, then the n-tuple

$$\mathbf{v}^{(1)} = v_{n-1} v_0 \cdots v_{n-2}$$

obtained by shifting v cyclically one place to the right is also a code word of C. From this definition, it is clear that

$$\mathbf{v}^{(i)} = v_{n-i} v_{n-i+1} \cdots v_{n-1} v_0 \cdots v_{n-i-1}$$

is obtained by shifting v to the right cyclically i places is also a code word of C.

Cyclic codes were first studied by E. Prange in 1957 [9]. Since then, progress in the study of cyclic codes has been improved upon by algebraic coding theorists. Cyclic codes are attractive for two reasons. First, encoding and syndrome calculation of a cyclic code can be easily implemented. Second, because they have nice algebraic structures, it is possible to find various efficient decoding methods. The binary (23, 12) Golay code is also a cyclic code. The cyclic structure of Hamming codes was studied by N. Abramson in 1960 [1].

Of the numerous classes of random error correcting codes, the class discovered by A. Hocquenghem (1908–1990) in 1959, and independently by R. C. Bose (1901–1987) and D. K. Ray-Chaudhuri (1933–) in 1960 is a large class of multiple error correcting codes, which are called the BCH codes. The BCH codes are also cyclic codes. Additional information on the BCH codes or other advanced error correcting codes can be found in any standard textbook on algebraic coding theory (e.g., [2, 3, 7]).

Since 1969, a number of advanced error correcting codes have also been developed for data transmission from deep space. Some of them are linear codes, and others are not. The spacecraft *Voyager 1* was launched in 1977 destined to explore Jupiter and Saturn. The extended binary Golay code was used as an encoder in the imaging system of the spacecraft [3]. A brief summary of the use of error correcting codes in the history of space exploration can be found in [3]. More thorough information about deep space applications of error correcting codes can be found in [13].

REFERENCES

1. N. Abramson, A Note on Single Error-Correcting Binary Codes, *IRE Transactions on Information Theory*, IT-6, 502–503, 1960.
2. Richard E. Blahut, *Theory and Practice of Error Control Codes*, Addison-Wesley, Reading, MA, 1984.
3. W. Cary Huffman and Vera Pless, *Fundamentals of Error-Correcting Codes*, Cambridge University Press, Cambridge, UK, 2003.
4. M. J. E. Golay, Notes on Digital Coding, *Proceedings of IRE*, 37, 657, 1949.
5. R. W. Hamming, Error Detecting and Error Correcting Codes, *Bell System Technical Journal*, 29, 147–160, 1950.
6. Z. Kiyasu, *Research and Development Data* (no. 4), Electrical Communications Laboratory, Nippon Television Network Corp., Tokyo, 1953.
7. F. J. MacWilliams and N. J. A. Sloane, *The Theory of Error-Correcting Codes*, Elsevier/North Holland, New York, 1977.
8. W. W. Peterson and E. J. Weldon Jr., *Error-Correcting Codes* (2nd ed.), MIT Press, Cambridge, MA, 1972.
9. E. Prange, *Cyclic Error-Correcting Codes in Two Symbols* (AFCRC-TN-57-103), Air Force Cambridge Research Center, Cambridge, MA, 1957.
10. D. A. Slepian, A Class of Binary Signaling Alphabets, *Bell System Technical Journal*, 35, 203–234, 1956.
11. A. Tietavainen, A Short Proof for the Nonexistence of Unknown Perfect Codes over GF(q), q > 2, *Annales Academiae Scientiarum Fennicae A*, 580, 1–6, 1974.
12. J. H. van Lint, A Survey of Perfect Codes, *Rocky Mountain Journal of Mathematics*, 5, 199–224, 1975.
13. S. B. Wicker, Deep Space Applications, in *Handbook of Coding Theory*, ed. V. S. Pless and W. C. Huffman, Elsevier, Amsterdam, 1998, pp. 2119–2169.
14. C. E. Shannon, A Mathematical Theory of Communication, *Bell System Technical Journal*, 27, 379–423, 1948.

15. C. E. Shannon, A Mathematical Theory of Communication (Part II), *Bell System Technical Journal*, 28, 623–656, 1949.
16. Wikipedia, Richard Hamming, http://en.wikipedia.org/wiki/Richard_Hamming.
17. D. E. Muller, Application of Boolean Algebra to Switching Circuit Design to Error Detection, *IRE Transactions on Electronic Computers*, 3, 6–12, 1954.
18. I. S. Reed, A Class of Multiple-Error-Correcting Codes and the Decoding Scheme, *IRE Transactions on Information Theory*, 4, 38–49, 1954.

CHAPTER 23

Automata and Formal Languages

The term *automaton* comes from the Greek word *automatos* meaning an "autonomous apparatus." In particular, an apparatus that looks like a human being or an animal was called an *automatos*. A contemporary robot can be considered an intelligent machine that has evolved through a long process from the ancient *automatos*.

23.1 AUTONOMOUS APPARATUS

The Greek scientist Heron Alexandria (c. 10–70) is a well-known inventor of various automata. He taught mathematics, physics, and mechanics at the Royal Museum of Alexandria. Some of his lecture notes still exist today. Some of Heron's autonomous apparatuses were operated by programmable procedures. These included the aeolipile (a heat-powered steam engine), a wind-powered musical instrument, automatic stage apparatus, a theater sound generator, a vending machine, and many others. The heavy doors of an abbey were opened and closed by Heron's steam engine. His vending machine dispensed holy water when coins were inserted. It is very surprising that such apparatuses were invented and operational almost 2000 years ago. The autonomous apparatuses invented by Heron are prototypes of the steam engines that appeared later, during the Industrial Revolution of the 18th century [4].

In the 12th century, during the golden age of the Islamic Empire, Al-Jazari (1136–1206), of northern Mesopotamia, also invented a number of autonomous apparatuses. Some of these included a device to pump water, a mechanical clock operated hydraulically, a human-like robot operated by programmable procedures, a device that converted rotational motion to reciprocal movement, and many others. These apparatuses are described in detail with his own illustrations in a book written in Arabic (1206). Its English translation, *The Book of Knowledge of Ingenious Mechanical Devices*, was published in 1973 [5].

The inventions and techniques of Heron and Al-Jazari eventually made their way to Europe. They were very influential in the development of mechanical instruments such as the early clocks of medieval Europe. Sketches of various advanced automata designed by Leonardo da Vinci (1452–1519) around the end of the 15th century have also been found. He hoped to build human-like automata capable of moving their heads and arms. Unfortunately, the technological level of medieval Europe was not sufficiently advanced to build such automata.

By the turn of the 18th century, various types of instruments and automata were being made in Europe. These included automata capable of playing musical instruments and automata that could write letters. Some of these automata remain intact even today. For example, at the entrances of arcades or main squares of cities in Switzerland, we can see old clocks animated by automata playing musical instruments and parading hourly. Most such clocks were built in prosperous cities of 18th- or 19th-century Europe. Various automata were also made at the same time in Japan. A doll-like automaton serving tea and a bird-like automaton carrying a written oracle are examples of Japanese-made automata from the 18th century.

23.2 AUTOMATA AS COMPUTING MODELS

Let us revisit the Turing machines described in Chapter 16. A Turing machine can be considered an abstract machine (i.e., an automaton) whose purpose is to compute a function, solve a problem, or recognize a language, which is a set of strings defined over some finite alphabet of symbols. Information or a stimulus given to the automaton from the outside is called an input action or simply an *input* to the automaton. During its computation, the status of the automaton normally changes step by step. We can consider that an automaton is a function from a sequence of inputs to a sequence of output symbols. If we are unconcerned about its physical construction, components used in the construction, or the material of the components, we can describe the automatic machine as a mathematical model. So Turing machines and actual machines, e.g., computers, can be viewed as special cases of automata.

What is the major difference between Turing machines and mathematical models of actual machines? Any physical machine assumes a state from a finite set of possible states. On the other hand, at any point in time during its computation, the *configuration* of a Turing machine is a pair consisting of its state (instruction being executed) and all of the contents of its work tape. In general, the number of different configurations taken by the Turing machine cannot be limited by a fixed number, since the number of its moves is unbounded. Therefore, the computational ability of Turing machines exceeds the abilities of any physical or any fixed-memory automata.

Quantum mechanical considerations aside, any automatic vending machine, elevator control system, and even the human brain can only be in one of a finite number of possible states. Therefore, we may consider automata, as models of actual machines, as really being restricted Turing machines that are not allowed to perform any write operations onto their work tapes. An automaton having a finite number of states is called a *finite state automaton* (FSA), a *finite state machine* (FSM), or simply a *finite automaton* (FA). Warren S. McCulloch (1898–1969) and Walter Pitts (1923–1969) introduced finite automata in 1943 as neural network models. Around the middle of the 1950s, David Huffman (1925–1999), Edward F. Moor (1925–2003), and George H. Mealy (1925–) defined a finite automaton in the form of a set of states, a state transition function, and an output function [6–8]. They demonstrated a number of fundamental properties of finite automata. Subsequently, many computer scientists and mathematicians showed interesting results about FAs.

FAs have many practical applications, e.g., in the design of logic circuits and control systems. Arithmetic and control circuits in a computer can be considered FAs. Suppose we want to design a logic circuit for a given Boolean function using as few gates as possible. To achieve this logic circuit, we first design the minimum state finite automaton such that its input-output function is equivalent to the given Boolean function. Then, we construct a logic circuit realizing the minimum state FA. As a simple example, let us consider an FA that accepts the language of all binary strings containing an odd number of 1's (i.e., an odd parity checker). Such a recognizer can be realized by a two-state FA, depicted in Figure 23.1. It should be obvious that we cannot design a one-state FA for this task.

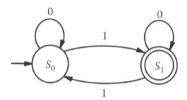

FIGURE 23.1   The state transition diagram of an odd parity checker FA.

The set of all possible input sequences/strings to an FA is grouped in two classes: those in the language of the FA, i.e., accepted by it, and the rejected ones, which are not in the language. The finite set of states in an FA is divided into two groups: *accepting* states and *nonaccepting* states. For a given input sequence, if the computation by an FA ends at an accepting state after the last input symbol has been read in, then the input sequence is said to have been *accepted*. That is, the input string is determined to be a member of the language for which the given FA was constructed. Otherwise, the input string is said to be *rejected*. In this way, each FA defines a set of input strings or sequences. If the number of states is not too big, an FA can be conveniently expressed as a state transition diagram, e.g., as shown in Figures 23.2 and 23.3. Otherwise, the FA's state transition function can be represented in a table.

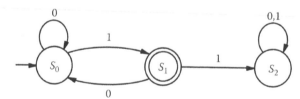

FIGURE 23.2   The state transition diagram of an FA recognizing strings not containing consecutive 1's.

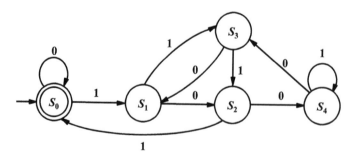

FIGURE 23.3   An FA recognizing binary strings of numbers divisible by 5.

The state transition diagram in Figure 23.2 defines an FA accepting the set of binary sequences not containing consecutive 1's. S_0, S_1, and S_2 are the three states of the FA. Its initial state, S_0, is indicated by the leftmost arrow, and the accepting state, S_2, is indicated by double circles. When any symbol, say a 0 or a 1, is received or read from the input tape, the FA state moves from its current state to the next, as defined by its *state transition function*, which is represented by the arrows marked 0 or 1, respectively. For example, if an input sequence of 00110 is supplied to the automaton in Figure 23.2, its state transition sequence would be S_0, S_0, S_0, S_1, S_2, S_2. Since S_2 is not an accepting state, this input sequence, which contains consecutive 1's, would be rejected.

The language recognized by an FA can be quite complex. Consider the set of binary strings representing numbers divisible by 5, for example. The five-state FA shown in Figure 23.3 is the smallest FA that correctly recognizes this language. Observe that this FA accepts the language of binary strings representing numbers *divisible* by 5, as opposed to strings whose *lengths* are divisible by 5, for which a different five-state FA can easily be constructed.

A set recognized by any FA can also be expressed or generated by a simple expression called a *regular expression*. The corresponding generated set is called a *regular set*. Regular expressions were introduced by Stephen C. Kleene (1909–1994) around the middle of the 1950s. The class of languages recognizable by FAs is the same as the class of regular sets. A regular expression over an alphabet (a finite set of symbols) can be constructed by using any of three types of operation symbols: union, concatenation, and star; any symbol of the alphabet; the symbol for the empty set, denoted by ϕ; and the symbol for the null length string, denoted by ϕ. More formally, regular expressions over an alphabet Σ are defined as follows [1]:

1. ϕ is a regular expression and denotes the empty set, i.e., a set with no strings in it.

2. ε is a regular expression and denotes the set $\{\varepsilon\}$, i.e., a set with a null length string.

3. For each a in Σ, a is a regular expression denoting the singleton set $\{a\}$.

4. If r and s are regular expressions denoting sets R and S, respectively, then $r + s$, rs, and r^* are regular expressions that denote the sets $R \cup S$, $R \circ S$, and R^*.

In the definition given above, $R \circ S$ is the concatenation of languages R and S, i.e., the set of strings formed by choosing any (prefix) string from the set R and following it with any (suffix) string from S. The * operation in R^*, sometimes called the *Kleene star*, is the set of strings formed by any number of concatenations of strings from R, including the null length string ε. For example, $(0 + 1)^*$ denotes the set of all possible binary strings, whereas $(0 + 1)(00 + 01 + 10 + 11)^*1$ denotes the language/set of all even length strings representing odd numbers.

The computational abilities of the FAs are severely limited. There are many important sets that cannot be recognized by any FA. That is, most nontrivial problems cannot be solved by FAs. For example, we can prove that there does not exist an FA that can decide whether a given input has the same number of 0's as 1's, or whether a given arithmetic expression is syntactically correct. An *arithmetic expression* can be expressed by properly combining any number of arithmetic operation symbols, such as +, –, ×, ÷, constants, variables, and parentheses. For example, $(a + b) \times (a - c)$ is a legitimate arithmetic expression, but $(a + (b - 1)$ is not. Note that the parentheses in $(a + (b - 1)$ are not being used properly. Therefore, FAs are not sufficiently powerful to recognize, for example, whether a computer program is *syntactically* correct. As a side note, Turing machines can easily make that determination; however, even they are not powerful enough to check for the *semantic*, or logical, correctness of computer programs.

If an auxiliary tape called a stack or pushdown is added to an FA, then the ability of the automaton is significantly enhanced. This class of automata is called *pushdown automata* (PDAs). A *stack* is a first-in-last-out data structure. That is, symbols/data may be entered or removed only at the top of the stack. When a symbol is entered at the top, the symbol previously at the top becomes second from the top. Similarly, when a symbol is removed from the top, the symbol previously second from the top becomes the top symbol. A stack of plates on a spring in a cafeteria has a similar structure. When the top plate is removed, the plate immediately below appears above the level of the counter. If a plate is put on top, the pile of the plates is pushed down. From this analogy, a stack automaton is also called a *pushdown*. We can construct a PDA that can decide whether a given arithmetic expression is in a correct syntactical form. PDAs were first introduced as a formal model by Anthony Oettinger (1929–) in 1961. Since then, PDAs have been studied by Marcel Paul Schützenberger (1920–1996), Sheila Greibach (1939–), Seymour Ginsburg (1928–2004), and many others. PDAs have many useful applications and have been used in many areas of computer science, including programming languages, compiler design, and query/natural language processing [1].

The computational ability of PDAs is superior to that of FAs, but inferior to Turing machines. Interestingly, Marvin Minsky (1927–) showed that PDAs with two stacks are just as powerful as Turing machines. This raises an interesting question: can we build automata that are more powerful than (one-stack) PDAs, but less powerful than Turing machines? The answer is yes. John R. Myhill (1923–1987) introduced yet another class of machines, called *linear bounded automata* (LBAs), which are Turing machines with a restriction on the size of their work tapes. LBAs can only use as much tape (memory) as the length of the input given to them. It can be shown that the computational power of LBAs is superior to that of the PDAs, but still inferior to the Turing machines.

There are numerous classes of automata whose computational power is between that of FA and Turing machines. Of these, PDAs and LBAs have been studied most extensively, not only because they were introduced first, but also because of many important theoretical as well as practical implications for the field of computer science in general. An interesting question regarding the computational power of these automata arises when one introduces *nondeterminism*, i.e., the ability of automata to *guess* or *choose* a correct computational path (when given such a choice), into their computations.

It turns out that nondeterminism does not make FAs or Turing machines any more powerful. However, the class of languages recognized by deterministic PDAs is a proper subset of the class accepted by nondeterministic PDAs. The language of *palindromes*, i.e., strings that read the same forward and backward, is the simplest example of the way in which nondeterminism enhances the computational power of PDAs. Although extensively studied, the question of whether nondeterminism adds computational power to LBAs is still open. Chapter 26 has an interesting discussion on whether or how nondeterminism could *speed up* computations for many practical problems in the areas of mathematical optimization, biology, chemistry, and economics, just to name a few.

23.3 FORMAL LANGUAGES

Avram Noam Chomsky (1928–) is an American linguist, philosopher, and cognitive scientist. Around the middle of the 1950s, Chomsky introduced the concept of a *generative grammar*, which is a set of production rules for constructing sentences in a given language. In other words, a generative grammar defines a language as a set of sentences or strings, which can be generated from an initial symbol by using the production rules of the grammar. The production rules in these grammars are in the form of $\alpha \to \beta$, where α and β are finite length sequences/strings, each defined over two finite sets of *terminal* and *nonterminal* symbols. The production rule $\alpha \to \beta$ of a grammar indicates that a substring α in a sequence can be replaced by β. The production rules of the grammar in Example 23.1 generate the language of well-formed sequences of parentheses. When constructing grammars, one needs to make sure the grammar satisfies two conditions: it generates every string in the language, and it does not generate any strings that are not part of the language under consideration.

Example 23.1

Let $S \to ()$, $S \to (S)$, $S \to SS$ be the three production rules in our grammar, which generates well-formed sequences of parentheses. We may also use an equivalent composite rule $S \to () \mid (S) \mid SS$ to describe this grammar. The set of nonterminal symbols is $\{S\}$, while the set of terminals is $\{(,)\}$. S is also the starting symbol of the grammar. Now, examine the string of terminals: $(()) () (())$. We can generate this string from S via the following sequence of production rule applications:

$$S \to SS \to SSS \to (S)SS \to (())SS \to (()) ()S \to (()) () (S) \to (()) () (()).$$

In the derivation above, as applied to the leftmost nonterminal, the third production rule $S \to SS$, the third production rule, the second production rule $S \to (S)$, the first production rule $S \to ()$, the second production rule, and the first production rule are used to obtain the final string from the initial symbol S.

Consider two additional languages and grammars that generate them, as shown in Example 23.2. Language L_1 consists of the strings that have the form $a^n b^n$, $n > 0$, and L_2 (a pseudocomplement language to L_1) has the strings $a^p b^q$, $p, q > 0$, $p \neq q$.

Example 23.2

The production rules for L_1 are $S \rightarrow ab \mid aSb$. For L_2 the production rules are $S \rightarrow A \mid B$, $A \rightarrow aab \mid aAb \mid aA$, $B \rightarrow abb \mid aBb \mid Bb$.

Based on the complexity of the production rules, Chomsky categorized generative grammars into four types, now called grammars of *Chomsky types 0, 1, 2, and 3*. This classification is called the *Chomsky hierarchy*, and the name also applies to the language classes, which are generated by their respective grammar types [2, 3].

- *Type 0* grammars have no restrictions on their production rules. They have the form of $\alpha \rightarrow \beta$, although α must contain at least one nonterminal symbol. Type 0 grammars are also called *unrestricted* or *phrase structure* grammars, and the languages generated by type 0 grammars are called *phrase structure languages*, or *recursively enumerable sets*. This language class is equivalent to the class of languages that can be computed/accepted by Turing machines. A surprising fact is that these languages are not closed under the operation of complementation; i.e., some languages may be generated by type 0 grammars, but their respective complementary languages may not have generative grammars.

- *Type 1* grammars are phrase structure grammars with an additional restriction on the production $\alpha \rightarrow \beta$ that the length of β is at least as long as that of α; i.e., the production rules are *length preserving* or *length increasing*. Type 1 grammars are usually called *context-sensitive grammars* (CSGs). The languages generated by these grammars are called *context-sensitive languages* (CSLs). The class of CSLs is closed under complementation.

- *Type 2* grammars are CSGs with an additional requirement that α should be a single nonterminal symbol. Type 2 grammars are often called *context-free grammars* (CFGs). The corresponding languages are called *context-free languages* (CFLs). The class of CFLs is not closed under complementation.

- *Type 3* grammars are CFGs with an additional restriction that all production rules are of the form $A \rightarrow wB$ or $A \rightarrow w$, where A and B are nonterminal symbols and w is terminal. Type 3 grammars are also called *right-linear* or *regular grammars*. The class of languages generated by type 3 grammars is equivalent to the class of regular sets. This class is closed under complementation.

Formal languages are defined by their generative grammars. Although Chomsky's linguistic theory started in the area of natural languages, generative grammars have been much more successful in the area of formal languages and high-level programming languages. Chomsky himself noticed that his generative grammars can be applied to programming languages.

Since the end of the 1950s, the relationships between generative grammars and automata have been studied extensively, and many important theoretical as well as practical results were obtained. Specifically, Chomsky and Miller showed the equivalence of regular grammars and regular sets in 1958. Chomsky showed the equivalence of phrase structure languages and recursively enumerable sets in 1959. Y. Bar-Hillel, M. Perles, and E. Shamir showed a number of interesting properties of context-free languages in 1961 [9]. S. Y. Kuroda showed the equivalence of linear bounded automata and context-sensitive grammars in 1964 [10].

Regular grammars play an important role in the specification of programming languages. Context-free grammars are important in the design of parsers for high-level programming languages, addressed in Chapter 25. Context-sensitive grammars have numerous applications in chemistry, medicine, and biological sciences.

REFERENCES

1. J. E. Hopcroft and J. D. Ullman, *Introduction to Automata Theory, Languages, and Computation*, Addison-Wesley, Reading, MA, 1979.
2. N. Chomsky, Three Models for the Description of Languages, *IRE Transactions on Information Theory*, 2, 113–124, 1956.
3. N. Chomsky, On Certain Formal Properties of Grammars, *Information and Control*, 2, 137–167, 1959.
4. Wikipedia, Hero Alexandria, http://en.wikipedia.org/wiki/Hero_Alexandria.
5. Al-Jazari, *The Book of Knowledge of Ingenious Mechanical Devices* (English trans.), Springer-Verlag, Berlin, 1973.
6. D. A. Huffman, The Synthesis of Sequential Switching Circuits, *Journal of the Franklin Institute*, 257, 161–190, 275–303, 1954.
7. G. H. Mealy, A Method for Synthesizing Sequential Circuits, *Bell System Technical Journal*, 34, 1045–1079, 1955.
8. E. F. Moore, Gedanken Experiments on Sequential Machines, in *Automata Studies*, Princeton University Press, Princeton, NJ, pp. 129–153, 1956.

9. Y. Bar-Hillel, M. Perles, and E. Shamir, On Formal Properties of Simple Phrase Structure Grammars, *Zeitschrift für Phonetik, Sprachwissenshaft, und Kommunikationsforrshung*, 14, 143–172, 1961.
10. S. Y. Kuroda, Classes of Languages and Linear Bounded Automata, *Information and Control*, 7, 207–223, 1964.

CHAPTER 24

Artificial Intelligence

Modern artificial intelligence (AI), as it is known today, emerged with the introduction of electronic computers (see Chapter 18) in the late 1940s [1, 2, 14, 15]. These machines were able to store and process information at very high speeds. This enabled researchers to build models and design solutions for complex problems in many problem domains, including those simulating human intelligence and "smart" behavior. Since the early years, progress in AI has been significant. This is true particularly in computer hardware, where the size and price of computers have shrunk while the speed and reliability have increased. On the other hand, the software AI advancements have been somewhat slower, but still many novel complex software tools and algorithms have been developed. However, AI progress there has not been as fast as initially expected. For example, it was predicted that computer chess programs would defeat human experts by the end of the 1960s. Only by 1997 did a computer program (*Deep Blue*) defeat Gary Kasparov, the human world chess champion [15]. Moreover, some AI experts consider that this victory was achieved only because of progress in hardware development, which made possible what was essentially a brute-force search, as opposed to a true algorithmic AI solution.

Starting in the middle of the 1980s *expert systems* started to easily outperform their human counterparts in fields such as medicine, the operation of complex distribution (e.g., water [3]) and control (e.g., system control [4]), and many other problem domains. While today's machines, automobiles, appliances, and almost all electronic devices seem to possess some form of an electronic "brain," these simple (control) devices should not be classified as part of AI per se. Human activities such as business or financial decision making, solving engineering problems, flying a plane, etc., require various levels of intelligence to be carried out. Computer systems that are smart, e.g., an autopilot, usually equal or exceed human abilities in their specific areas of expertise. Any software or hardware system that can successfully perform such nontrivial tasks is said to possess some degree of artificial intelligence. While some experts believe that artificial intelligence of computers will eventually surpass the natural intelligence of humans, it is the opinion of most that that will not happen in the foreseeable future.

The one thing the majority of experts do agree on is that performing human tasks will require an extremely complex set of AI processes, which must then be designed into algorithms that will eventually be implemented as AI software. In this chapter, we discuss the history of artificial intelligence, including its most important milestones, as well as the scientists whose pioneering work has made AI what it is today.

24.1 WHAT IS AI?

There are many definitions and explanations of what artificial intelligence is (or is not). Below, we list some of them together with the name of the person who originated it and the year it was originally introduced:

1. "The science and engineering of making intelligent machines" (John McCarthy (1927 –), 1956, e.g., see [11]).

2. "Machine intelligence is an enterprise which may eventually offer yet one more mirror for man, in the form of a mathematical model of knowledge and reasoning" (Donald Michie (1923–2007), 1973, e.g., see [12]).

3. "The automation of activities that we associate with human thinking—activities such as decision-making, problem solving, and learning, i.e., the characteristics we associate with intelligence in human behavior and that require common sense and intelligence when performed by people" (a compilation of Bellman, Feigenbaum, and Kurzweil).

A very common definition of *artificial intelligence* is that it is the creation of virtual machines "who can think." While AI is clearly a field of computer science, more often than not, it addresses problem domains that are outside of CS; i.e., it deals with the creation and development of intelligent machines/systems, methodologies for performing tasks that require human intelligence in general, and in particular, the computerized automation of intelligent behavior.

24.2 AI TIMELINE

As described in several of the previous chapters, one could say that the historical roots of AI extend for thousands of years. Various artifacts and then mechanical or computational devices with some level of intelligence had been invented as early as 3000 BC, although one does not seriously consider pebbles or the counting stick as intelligent. It was only after modern electronic computers were developed (see Chapter 18) that AI "took off" and became what it is today. While all programs perform tasks for which they were designed, today one would not say that adding two numbers is an intelligent task.

Below is a timeline (by no means exhaustive) of some of the more important milestones, not necessarily computer based, in the history of AI. This list was compiled from several sources, including [1, 2, 8, 14, 15]. Also note that almost all of the entries enumerated below are discussed in detail in various chapters throughout this book:

1. The first "surgical expert system" was introduced around 3000 BC. This system (written on papyrus) was based on 48 head wound surgical observations, which were compiled using specific symptom-diagnosis and treatment-prognosis combinations using a simple "If ..., then ..." expert system approach.

2. In the 13th century, the Zairja was invented by Ramon Llull (c. 1232–c. 1315). It was a device that was to systematically generate ideas by mechanical means.

3. An eight-digit calculator, called the Pascaline, was invented by Pascal in 1642.

4. Gottfried Leibniz constructed his calculator in 1694. His algorithm is still in use today.

5. In 1726, Jonathan Swift (1667–1745), in his famous book *Gulliver's Travels*, predicted the invention of an automatic book writer.

6. The first truly programmable device was invented by Joseph Jacquard (1752–1834) in 1805. His device was a programmable loom with instructions/patterns provided on punched cards.

7. The analytical engine was designed by Charles Babbage (1791–1871) in 1832. This engine was the first mechanical programmable computer. Although the original never truly worked, a model built in 1991 (using Babbage's plans) did.

8. A chess automaton, which played chess end games without human intervention, was constructed by Leonardo Torres (1852–1936) in 1910.

9. The term *robot* was invented by the Capek brothers (Josef Capek, 1887–1945; Karel Capek, 1890–1938) in the 1920s, with Josef actually coining this word. It is derived from the Czech-Polish word *robota*, meaning "work." In Karel's 1923 play *Rossum's Universal Robots*, the term *robot* was used to describe intelligent machines that revolted against their human masters.

10. In 1928, John von Neumann came up with the minimax theorem, which is still being used by a majority of game-playing programs (e.g., in chess).

11. The Turing machine, which can carry out the operation of any other computing machine or a well-defined procedure, was introduced by Alan Turing in 1937.

12. In 1938, Claude Shannon (1916–2001) showed that calculations could be performed much faster using electromagnetic relays than with mechanical calculators. The electromagnetic relays were used in the world's first operational computer, Heath Robinson, in 1940, built by the English to decode the German Enigma messages.

13. In the mid-1940s, the electromechanical relays in calculators were replaced by vacuum tubes. This technology was applied on the Colossus [5] computer that was also used to decipher the most complex of German codes during World War II.

14. In 1945, John von Neumann introduced a basic computer architecture design that is still in use today.

15. The first general purpose, fully electronic, programmable computer, Electronic Numerical Integrator and Calculator [6] (ENIAC), was built in 1945. ENIAC was 1000 times faster than the electromechanical computers.

16. Circa 1946, *intelligence* was determined to be the process of processing information to achieve goals.

17. Learning artificial neural networks were proposed by Donald O. Hebb (1904–1985) in 1949.

18. The three laws of robotics were proposed by Isaac Asimov (1920–1992) in 1950.

19. The now-famous Turing test [13] was proposed by Alan Turing in 1950. Its purpose was to recognize machine intelligence. Turing argued that the machine should be considered intelligent if it could successfully pretend to be human to a knowledgeable observer and not be recognized as pretending to be such. Any machine passing the Turing test should be considered intelligent.

20. The first artificial neural network was built by Marvin Minsky (1927–) [10] and Dean Edmonds in 1951. It simulated a rat trying to find its way through a maze.

21. In 1955, Allen Newell (1927–1992), Herbert Simon (1916–2001), and J. C. Shaw wrote an AI program, called the Logic Theorist. It could be used to prove theorems using a combination of searching, goal-oriented behavior, and application of logical rules. The program was written in a new computer language called Information Processing Language.

22. In 1955, the term *artificial intelligence* was coined by John McCarthy (1927–) in his proposal for a conference at Dartmouth [11]. In the summer of 1956, a 2-month conference/study at Dartmouth was attended by many scientists, who today are considered to be the pioneers of artificial intelligence. The participants included John McCarthy, Marvin Minsky, Nathaniel Rochester (1919–2001), Claude Shannon, Herbert Simon, and Allen Newell. The Dartmouth conference was to proceed "on the basis of the conjecture that every aspect of learning or any other feature of intelligence can in principle be so precisely described that a machine can be made to simulate it" [2, 11]. The two most significant outcomes of the Dartmouth conference were a new paradigm of symbolic information processing, which was introduced by Allen Newell and Herbert Simon in their Logic Theorist computer program, and the introduction of the term *symbolic AI*, as opposed to *brain modeling*, which until that time had been used by the majority of AI researchers.

23. In 1957, Newell, Shaw, and Simon came up with their General Problem Solver, the first computer program that kept its knowledge of problems (rules represented as input data) apart from its strategy of how to solve them.

24. The Geometry Theorem was written by Herbert Gelernter in 1957. This computer program used a three-step proof of the geometry theorem to prune a search with a billion alternatives down to only 25 alternatives.

25. Also in 1957, Arthur Samuel (1901–1990) wrote a checkers program, which later learned how to beat Samuel—the one who wrote the program.

26. The Artificial Intelligence Laboratory at the Massachusetts Institute of Technology was founded by John McCarthy and Marvin Minsky [10] in 1958. Also, McCarthy developed the LISP programming language for AI applications.

27. In 1961, an anti-AI book was written by Mortimer Taube (1910–1965) entitled *Computers and Common Sense: The Myth of Thinking Machines*.

28. The first industrial robots were marketed by a U.S. company in 1962.

29. From 1965 to 1975, the first expert system, DENDRAL, was built at Stanford by Edward Feigenbaum (1936–) and Robert Lindsay, mapping the structure of complex organic chemicals.

30. The Blocks Microworld Project was created by Marvin Minsky and Seymour Papert (1928–) at the MIT AI Laboratory in the late 1960s. The goal was to integrate and improve computer vision, robotics, and natural language processing. This allowed computers to view and intelligently manipulate a simple "world of blocks" of various colors, shapes, and sizes under different physical configurations.

31. In 1969, a mobile robot called Shakey was built at Stanford. It was able to navigate a block world in eight rooms, after being given oral instructions.

32. In the early 1970s, Abe Mamdani and Seto Assilian used fuzzy logic [7] to control the operation of a small steam engine at the Queen Mary College in London. This was the first practical demonstration of the use of fuzzy logic for process control.

33. In 1972, the computer language PROLOG (Programmation en Logique) was developed by Alain Colmerauer and Philippe Roussel.

It became Europe's AI applications language of choice, while within the United States, LISP [9] was preferred.

34. In 1972, Stanford's Edward Shortliffe (1947–) wrote the MYCIN expert system, the world's first nontrivial medical software tool that diagnosed infections and blood diseases and recommended antibiotics, with dosages adjusted for patients' body weight.

35. Freddy (1969–1972) and Freddy II (1973–1976) are robots built by Donald Michie's group at the University of Edinburgh. They used computer vision to locate and assemble objects.

36. John McDermott built the first commercial expert system at CMU in the late 1970s. The system, XCON, was developed for Digital Equipment Company, which started using it in 1980 to automatically configure VAX computer systems.

37. Herbert Simon's theory of bounded rationality and satisfying behavior, considered to be one of the cornerstones of AI, was the major contributing factor in him winning the Nobel Prize in Economics in 1978.

38. In 1979, Hans Moravec (1948–) designed and built the Stanford Cart, the first car fully controlled by a computer.

39. In the 1980s, using fuzzy logic [7], Hitachi built a predictive control system for the automated subway trains in Sendai, Japan.

40. In 1984, GE built the Diesel Electric Locomotive Troubleshooting Aid, an expert system to diagnose breakdowns and give specific repair instructions.

41. In 1985, speech recognition systems started to work, providing continuous, speaker-independent speech recognition.

42. Polly, the first robot using computer vision and behaving as an animal, was built by Ian Horswill in 1993.

43. In 1994, Ernst Dickmanns (1936–) and Daimler-Benz ran their twin robot cars, VAMP and VITA2, for more than 1000 kilometers on a Paris highway.

44. More than 40 teams of robotic soccer players competed in the first RoboCup competition in 1997.

45. In 1999, the Sony Corporation introduced its robotic pet dog, AIBO (Artificial Intelligence Robot). It understood over 100 voice commands, used computer vision to see its environment, and had adaptive learning capabilities.

46. In 2002, a vacuum cleaning robot called Roomba was built. Two million were sold by 2006.

47. The first DARPA Grand Challenge was sponsored by the Defense Advanced Research Projects Agency (DARPA) in 2004. The challenge was a competition for autonomous (driverless) vehicles, and it took place on a desert course; however, no vehicles completed the course. The second DARPA Grand Challenge took place in 2005. Out of 23 vehicles, 5 completed the course. Stanley, an entry from Stanford, was the winner.

48. An artificial intelligence humanoid robot, ASIMO, was introduced by Honda in 2005. It walked as fast as humans and delivered trays in a restaurant.

49. In July 14, 2006, the Dartmouth Artificial Intelligence Conference: The Next 50 Years took place.

50. Watson, built in 2011 by IBM's David Ferrucci et al., won $1 million by beating human knowledge experts in the *Jeopardy* TV show. Winning the game required a deep understanding of natural language semantics and universal knowledge of facts.

24.3 AI PIONEERS

Many scientists have made significant contributions to the field of modern artificial intelligence. The pioneering work of the individuals discussed below has had a deep-seated impact not only on AI, but also, in most cases, on other areas of computer science, mathematics, and related sciences.

Alan Turing (1912–1954) was an English mathematician, logician, cryptanalyst, and computer scientist. One might argue that besides being the *Father of Computer Science*, Turing's contributions to artificial intelligence were, by themselves, fundamental. In 1936 Turing published his paper on computable numbers, in which the *Turing machine* was introduced. One of the most remarkable features of Turing's work, especially on his machine, was that he described modern computers even before they existed. Turing presented his universal machine (the universal Turing machine) as the one "which can be made to do the work of any special-purpose machine, that is to say to carry out any piece of computing, if a tape bearing suitable 'instructions' is inserted into it" [13]. Hence, Turing's universal machine was a computer and the instruction tape was a program that was run by that computer. In 1950, Turing published his paper "Computing Machinery and Intelligence in Mind" (see also Chapter 16), where he studied the problems that today are at the heart of artificial intelligence. In this paper, Turing introduced the *Turing test*, through which the intelligence of a computer can be determined.

Claude Shannon (1916–2001) is considered to be the Father of Information Theory. Shannon received his master's and doctorate from MIT in 1940, where he studied electrical engineering and mathematics. For his master's degree, which was in electrical engineering, Shannon applied George Boole's logical algebra to the problem of electrical switching. For his doctorate, Shannon applied mathematics to genetics. Besides his pioneering work on the theory of communication, Shannon wrote a paper "A Symbolic Analysis of Relay and Switching Circuits," where he pointed out the similarity between the truth values of symbolic logic and the binary values 1 and 0 of electronic circuits. He showed how switching circuits can be used to build a logic machine corresponding to the propositions of Boolean algebra. After Shannon joined Bell Telephone Laboratories as a research mathematician, he worked on the problem of most efficiently transmitting information. There, Shannon also showed the similarity between Boolean algebra and telephone switching circuits. During World War II, Shannon worked with Alan Turing, who had spent several months with Shannon while visiting the United States. They were interested in the possibility of building a machine that could imitate the human brain. They also worked together to build an encrypted voice phone that would allow President Roosevelt to have a secure transatlantic conversation with Prime Minister Churchill. *Information theory* was first introduced by Claude Shannon in his paper "A Mathematical Theory of Communication" in 1948. It progressed from a single theoretical paper to become a broad field. Shannon explained the way of quantifying information with absolute precision. He showed the essential unity of all information media. Shannon showed that every mode of communication, such as telephone signals, text, television and radio waves, and most other data, can be encoded in *bits*, a term he coined (see also Chapters 15 and 21). In 1950, Shannon wrote a groundbreaking paper on computer chess entitled "Programming a Computer for Playing Chess," in which he showed how a computer could play the game intelligently. His approach for having the computer program decide on the best move in a given chess game position incorporated John von Neumann's *minimax* procedure.

Nathaniel Rochester (1919–2001) received his B.S. degree in electrical engineering from MIT in 1941. After his graduation, Rochester continued at MIT working in the Radiation Laboratory. In 1948, he moved to IBM and designed the first general purpose mass-produced computer, IBM 701, for which he wrote the first symbolic assembler. In 1948, a group led by Rochester simulated the behavior of abstract neural networks on a computer. Rochester worked with Claude Shannon et al. to convince the Rockefeller Foundation to sponsor the Dartmouth conference in 1956, widely considered the "birth of artificial intelligence." He was a supervisor for many artificial intelligence projects at MIT, including Arthur Samuel's checker program, Herbert Gelernter's Geometry Theorem, and Alex Bernstein's chess program. In 1958, he worked with John McCarthy in the development of the LISP programming language, which was one of the first high-level languages designed specifically for AI.

John McCarthy (1927–2011) received his Ph.D. in mathematics from Princeton University. In 1948, McCarthy heard John von Neumann talking about self-replicating automata, which are machines that can create copies of themselves. McCarthy's research interests then turned to studying the relationship between human intelligence and machine intelligence. After the 1956 Dartmouth conference, McCarthy started working on computers that can play games and do other human-like tasks. McCarthy was the principal developer of the programming language LISP for AI applications. LISP is still in wide use today, especially in the United States. In 1990, McCarthy published his book *Formalizing Common Sense*, which contains a very nice compilation of his AI research results (see also Chapter 25).

Marvin Minsky was born in 1927. In 1950, Minsky received his B.A. from Harvard University, and in 1954 a doctorate in mathematics from Princeton University. In 1951, Minsky built the first neural network simulator and the first randomly wired neural network learning machine. His work laid down the foundations for the analysis of neural networks. Minsky built mechanical arms, hands, and other robotic devices. Minsky also did pioneering work on knowledge representation, theory of frames, and other well-known AI models. In the early 1970s, while at MIT, Minsky and Seymour Papert developed a theory, called the society of mind, which attempted to explain how a product of the interaction of nonintelligent parts could, itself, become intelligent. In the 1980s, Minsky showed how neural networks could be generated automatically, or self-replicate, in accordance with any arbitrary learning program, hence allowing *artificial brains* to be grown by a process similar to the development of a human brain.

24.4 AREAS OF AI

The field of artificial intelligence is not clearly partitioned into specific sub-areas or branches. For most applications, it is not unusual that several of them overlap at various levels. The following are some of the better-known and established areas of AI:

Game playing. There are programs that can easily defeat 99.9% of the best human chess players, including those at the grandmaster level or even the world chess champions (as happened in 1997). Usually, such programs take full advantage of the brute-force-type approaches that allow them to consider millions of moves per second and choose the best (optimal) one. While for almost any game there are computer programs that will beat the best of human players, the ancient Chinese game of Go still has human experts winning consistently (for now).

Speech recognition. Speech recognition techniques are used for specific purposes, such as flight numbers and city names in airline reservation systems. Many users, however, believe that human interaction or using computers with keyboard and mouse is still more convenient and reliable than using speech recognition computers.

Natural language processing. AI researchers believe that providing computers with sequence of words or syntactically parsing sentences is not sufficient. They believe that computers must be able to understand the domain, or the semantic context, that the text belongs to. For this reason, this ability is possible only for very limited domains. For similar reasons, *data mining*, in which semantic information is extracted from huge amounts of (medical, biological, geographical, etc.) data, still does not have a universal (domain-independent) approach for extracting information.

Computer vision. To make a computer vision program work perfectly, the information should be provided in three dimensions. However, most vision programs work on two-dimensional views. There are some programs that can analyze three-dimensional information, but they are not as good as the human eyes.

Expert systems. In expert systems, the knowledge that experts have is used to design programs that can do the tasks that these experts do. MYCIN was one of the first expert systems. It was designed in 1974 to diagnose bacterial infections of the blood and suggest treatments. MYCIN outperformed medical students and practicing doctors.

Logical AI. Here the programs are aware of the facts of any specific situation and they act in order to meet a specified goal. Such goals are usually represented by sentences of mathematical logical language (e.g., in PROLOG). The programs decide what to do by inferring which actions are appropriate for achieving the specified goals.

Fuzzy logic. Invented by Lotif Zadeh (1921–), it is a multivalued logic derived from fuzzy set theory. It deals with approximate, as opposed to precise, reasoning. The variables of fuzzy logic may have truth values between 0 and 1 (i.e., they are not Boolean). Specific functions are used to manage these variables. It has been applied to many areas of AI and others (e.g., control theory).

Search. Since AI programs examine a large number of possibilities, such as moves in a chess game, the goal of the search is to find a specific state efficiently among various domains. Some types of search techniques used in AI include *heuristic search, hill climbing search, best first search,* and *depth first search*. Various search space reduction techniques, such as alpha-beta pruning, are often employed to further reduce the tree of possibilities for certain (e.g., in minimax) applications.

Pattern recognition. Here, programs compare what they see with some predefined pattern. For example, a program can try to match eyes, mouth, and nose in order to recognize a face. Also, programs that deal with natural language text or chess position use more complex patterns.

Representation. In this branch of AI, languages of mathematical logic are used to represent facts about the world in some way.

Inference. Inference AI works as follows: a conclusion is to be inferred by default, but the conclusion can be withdrawn if there is evidence to the contrary.

Commonsense knowledge and reasoning. This area of AI has been an active research area since the 1950s. It is the farthest area from the human level in the field of AI, and more ideas are still needed to develop it.

REFERENCES

1. Wikipedia, Artificial Intelligence, http://en.wikipedia.org/wiki/Artificial_intelligence.
2. A Brief History of Artificial Intelligence, AtariArchives.org—Archiving Vintage Computer Books, Informaton and Software, http://www.atariar-chives.org/deli/artificial_intelligence.php.
3. P. Boulos, T. Altman, F. Collevati, and P. Jarrige, A Discrete Simulation Approach for Network Water Quality Models, *ASCE Journal of Water Resources Planning and Management*, 121(1), 49–60, 1995.
4. T. Altman, T. Hughes, and A. Wala, Mine Ventilation Expert System, *Applied Artificial Intelligence: An International Journal*, 2, 265–276, 1988.
5. Wikipedia, Colossus Computer, http://en.wikipedia.org/wiki/Colossus_computer.
6. ENIAC Computer History—Invention of the ENIAC Computer, The Great Idea Finder—Celebrating the Spirit of Innovation, 1997, http://www.idea-finder.com/history/inventions/comeniac.htm.
7. Wikipedia, Fuzzy Logic, http://en.wikipedia.org/wiki/Fuzzy_logic.
8. N. Hardman, The History of Artificial Intelligence, Docstoc—Documents, Templates, Forms, Ebooks, Papers and Presentations, http://www.docstoc.com/docs/2201141/The-History-of-Artificial-Intelligence.
9. Wikipedia, Lisp (Programming Language), http://en.wikipedia.org/wiki/Lisp_(programming_language).
10. Wikipedia, M. Minsky, http://en.wikipedia.org/wiki/Marvin_Minsky.
11. J. McCarthy, A Proposal for the Dartmouth Summer Research Projection Project on Artificial Intelligence, Formal Reasoning Group, 1955, http://www-formal.stanford.edu/jmc/history/dartmouth/dartmouth.html.

12. D. Michie, *On Machine Intelligence*, Edinburgh University Press, Edinburgh, 1973.
13. J. J. O'Connor and E. F. Robertson, Alan Mathison Turing, MacTutor History of Mathematics, School of Mathematics and Statistics, University of St. Andrews, Scotland, 2003, http://www-history.mcs.st-and.ac.uk/Biographies/Turing.html.
14. H. Stottler, Smarter Software Solutions, http://www.stottlerhenke.com/ai_general/history.htm.
15. Wikipedia, Timeline of Artificial Intelligence, http://en.wikipedia.org/wiki/Timeline_of_artificial_intelligence.

CHAPTER 25

Programming Languages

This chapter touches on a number of influential programming languages and also discusses the characteristics of several categories of programming paradigms. These paradigms include imperative, declarative, functional, logical, and object-oriented programming.

While certain languages are commonly known by a particular programming paradigm—Fortran is known as an imperative language, LISP as a functional language—it is often the case that languages are hybrids to a certain degree, combining aspects of several paradigms into their specifications. As intricate detail is not the goal of this chapter, subtleties such as these are not always addressed.

25.1 MACHINE CODE

The first programming languages implemented in the 1940s and early 1950s were slow, difficult to use, unfriendly, and error-prone, much like the first computers on which they ran. At the time these languages were developed and used, they were known as pseudocodes, though the definition of the word has evolved significantly since that time [1]. Pseudocodes were not high-level languages, nor even assembly languages; they were bare-bones machine languages. Programs were exceedingly difficult to write, difficult to read, difficult to maintain, and incredibly fragile. Numeric codes represented instructions, absolute addressing meant that inserting a single line could invalidate every address call that followed, and computer architectures were not capable of floating-point arithmetic or indexing for arrays [1].

These and other shortcomings inherent in machine language and in simple computer architectures motivated the development of a more advanced and abstracted programming tool: the assemblers and assembly languages that evolved during the early 1950s [1]. Although assembly code was a notable improvement from machine code, assembly languages did not have much impact on the development of higher-level languages.

25.2 INTERPRETATIVE CRUTCHES

In the 1950s, several interpretive systems were developed that extended machine code to allow for floating-point operations. John Backus (1924–2007) developed one such system called the Speedcoding system for the IBM 701. The interpreter was able to use the architecture to represent a virtual three-address floating-point calculator. Instructions were developed to execute addition, subtraction, multiplication, division, square root, sine, arc tangent, exponent, and logarithm functions. Speedcoding also allowed for conditional and unconditional branches and I/O conversions as part of its virtual architecture. Speedcoding even automatically incremented the address registers when a new line of code was added (though not until 1962). Problems that took days or even weeks to program in machine code could be programmed in a few hours using Speedcoding. However, the remaining usable memory after loading the interpreter was generally very small, and instructions took a significant amount of time to execute because simulating floating-point operations in software was a very time-consuming process [1].

25.3 THE FIRST HIGH-LEVEL LANGUAGE: FORTRAN

Today's high-level languages are abstract, flexible, and portable; they are able to be written, compiled, and run on virtually any modern computer. In the early days of computing, the situation was quite different; programming features were dependent on the architecture of the particular model of machine targeted. So was the state of technology during the groundbreaking development of the first high-level language.

The IBM 704 was the first mass-produced computer with floating-point arithmetic hardware. Including in hardware what until then could only be emulated by interpretive systems, it provided the foundation needed for the next big step in computing. The 704 was released by IBM in 1954, and marked the beginning of the end for the slow, memory-intensive interpretive systems. Making good use of the advanced new hardware, the first widely accepted high-level programming language was developed. It was named the IBM Mathematical Formula Translating System, or Fortran.

Although a game-changing advancement, Fortran, like its more basic predecessors, was still largely a product of its environment. Even with the fancy new 704, computers were very slow and suffered from unreliability and cripplingly small memories. Compiled code speed was the primary goal of the first Fortran compilers. Since computers were used primarily for scientific computing, there were no existing effective or efficient ways to program computers, and hardware was very expensive.

Fortran 0 was the initial version of Fortran and was fully described before the implementation of the language began. Subsequent versions of Fortran were I (1957), II (1958), III (1958), IV (1960), 77, 90, 2003, and 2008 [1].

Because Fortran was the first high-level language, it had a dramatic effect on shaping the way computers were used and the way other high-level languages were developed. Fortran's original designers never intended it to run on anything but an IBM machine, nor to target any field other than numerical computation. While early versions were certainly lacking in many ways, this is to be expected during the evolution of the first high-level language. Overall, the language and its influence on programming were a monumental success; Fortran pioneered a concept that influenced all else that followed.

Just as new Fords have rolled off the production line since the first Model T's in 1910, Fortran has evolved a great deal since its conception. It continues to be used today as a general purpose, procedural, imperative programming language. As was the intent of its designers, it is still well suited for numeric and scientific computation.

25.4 OVERVIEW: IMPERATIVE PROGRAMMING

Imperative programming is a programming paradigm in which explicit statements or commands are used to change the state of a program. Imperative programs define sequences of statements or commands to be performed by a machine in order to get into a particular state. Imperative programming is the polar opposite of declarative programming.

Procedural programming is a flavor of imperative programming in which the program is built from procedures or functions. In procedural programming, state changes are localized to procedures or restrained by the parameters passed in to and return values passed out of procedures. It is known that this technique is useful for structured programming.

While structured programming is possible to an extent in any programming language, it fits best with procedural programming languages.

ALGOL, Pascal, PL/I, and Ada are examples of some early imperative, procedural languages where structured programming was commonly used. In addition to Fortran, C is also an imperative, procedural language.

25.5 OVERVIEW: DECLARATIVE PROGRAMMING

Declarative programming is a programming paradigm in which the program describes its desired results without explicitly specifying a list of commands to execute in order to achieve those results. Functional and logical programming languages are generally declarative in style.

25.6 THE SECOND HIGH-LEVEL LANGUAGE: LISP

Around the time that Fortran II was released, a second high-level language was designed and implemented: LISP. It was invented by John McCarthy in 1958 while he was at MIT [3]. LISP (from List Processing) was the first functional programming language and was developed to provide language features for list processing, which was needed by the first artificial intelligence (AI) applications of that time (See Chapter 24).

At a time when most computations were done on numeric data in arrays by Fortran, developers needed a method to allow computers to process symbolic data in linked lists. Unlike scientific computations and number crunching, AI was a field where it was often not straightforward to specify the problem to be solved or the method of solving it, and success in this area required a much different approach.

LISP was not the first attempt at a list processing language. A language named Information Processing Language I (IPL-I) was described as a theoretical list processing language, and was implemented by version II. IPL lived on until its fifth version, but was never widely used due to its low-level syntax; it was an assembly language implemented with an interpreter that could handle list processing instructions. Furthermore, IPL was implemented on and for a RAND Johnniac computer, which was a long-lived but obscure machine [1].

In the mid-1950s IBM decided to implement a list processing language based on Fortran, so as to reuse the existing (expensive!) Fortran I compiler. The language was called the Fortran List Processing Language (FLPL) and was an extension to Fortran. It was used to prove plane geometry theorems, a relatively easy area for mechanical theorem proving [1]. However, FLPL's shortcomings—no support for recursion, dynamic storage allocation, or conditional expressions—led to the development of the more capable LISP.

LISP addressed these shortcomings of Fortran and pioneered several other new advancements, which include automatic storage management through implicit deallocation, dynamic typing, and the ability of a LISP compiler to compile its own source code.

The initially developed language is now known as pure LISP, because it is a purely functional language. Common LISP is one well-known descendant of pure LISP, as is Scheme. Common LISP supports some procedural and some object-oriented programming capabilities in addition to its inherited functional features, and was developed to provide a standard for several dialects of LISP that suffered from portability issues. Scheme uses typed variables and treats functions as first-class entities, meaning that functions can be passed as arguments to other functions or assigned to variables as values. LISP, Common LISP, and Scheme are still in use today, though more heavily in academia and theory/research than in industry.

25.7 OVERVIEW: FUNCTIONAL PROGRAMMING

In *functional programming*, computation is accomplished by applying mathematical functions to arguments. Functional programming avoids the concepts of program state and state changes that are present in imperative programming. The variables and assignment statements of imperative programming are also absent (and unnecessary) in functional programming. Loops are also unnecessary in functional programming because functions can be called recursively.

These basic differences make programming in a functional language very different from programming in an imperative language. Imperative functions can have side effects that can change the program's state. The same expression can result in different values at different times. The opposite is true in functional programming, where the value of a function depends only on its arguments. Thus, a function will always produce the same results if given the same arguments since there is no program state. This can make functional programming more straightforward when attempting to understand, predict, and troubleshoot programs.

25.8 STANDARDIZATION AND COMPROMISE: ALGOL 60

In addition to Fortran and LISP, there were several other high-level languages in the works in the late 1950s. The boom of new languages led to troubles in sharing programs among different users and platforms;

most of the new languages were geared toward a single architecture. It quickly became obvious that machine-dependent languages were not the way to foster widespread collaboration, portability, and standardization. In response to this growing problem, several large computer user groups in the United States banded together to submit a petition to the Association for Computing Machinery (ACM) in 1957 to form a committee to spearhead the creation of a machine-independent scientific programming language [1]. While capable, Fortran was not a practical candidate for this universal programming language since it was owned exclusively by IBM at the time.

A few years before this, another group in Germany called the Society for Applied Mathematics and Mechanics had convened with the same purpose in mind: to design a universal programming language. In 1958, these two groups joined forces and began a collaborative language design project [1].

The goals laid out in the first week-long design meeting for the new universal language were [1]:

- The syntax of the language should be as close as possible to standard mathematical notation to make programs highly readable with little to no further explanation needed.

- The language used for the description of algorithms should be able to be used in printed publications.

- Programs in the language must be mechanically translatable into machine language (a requirement for any programming language).

The language developed at that design meeting met these goals, but the design process required a great deal of compromise from everyone involved. The resultant language was originally named the International Algorithms Language (IAL), but was later changed to ALGOL (Algorithmic Language), and then settled as ALGOL 58.

ALGOL 58 was similar to Fortran in many ways, which is logical considering that the primary concern of both languages was scientific programming. To free the language from being tied to a particular machine and to make it more flexible and capable, many of the features available in Fortran were generalized in ALGOL 58. Several new ideas were included as well, such as formalizing the concept of a data type and allowing compound statements. The reception to the new language was enthusiastic.

In early 1960, the members of the second ALGOL design meeting discussed how ALGOL 58 should evolve. As an interesting side note, Backus had introduced his new language description syntax, Backus–Naur Form (BNF), the year before, and Naur wrote a description of the newly proposed language in BNF and distributed it to all of the members at the beginning of the meeting. By the end of the 6-day-long conference, the new and improved ALGOL 60 had been specified. The scope and magnitude of the agreed-upon changes were significant. Some of the more important new developments included:

- The block structure, which introduced nested scopes

- The ability to pass parameters by either value or name/reference

- Recursive procedures (old hat for LISP, but new for imperative programming)

- Stack-dynamic arrays (the size of and storage for an array is allocated during execution)

Although ALGOL 60 was used mostly by computer scientists in the United States and Europe, it was never widely used in commercial applications [4]. However, most imperative programming languages designed since that time descend from ALGOL 60 either directly or indirectly. These languages include PL/I, SIMULA 67, ALGOL 68, C, Pascal, Ada, C++, Java, and C# [1].

Some of the factors that held ALGOL 60 back were its lack of input and output statements. Some of its features were too flexible to understand and implement well, and the support to ALGOL 60 from IBM was not sufficient. For these reasons, it could not compete with the existence of popular Fortran in numerical computation applications.

25.9 FROM SCIENCE TO BUSINESS: COBOL

COBOL is an interesting case study in programming. The language was widely used in the business world—perhaps the most widely used, at least near the end of the last millennium—but it has had little to no effect on subsequent languages (with the exception of PL/I). Like ALGOL 60, COBOL was designed by a committee in 1959, but in this case by a committee sponsored by the U.S. Department of Defense (DoD). One of COBOL's design goals was that it should use English as much as possible and be easy to use, even at the expense of power, in order to open up development to as many people as possible. It was also decided that the design should not be restricted by problems of implementation, as had happened in several areas with ALGOL.

No doubt because of its sponsorship, COBOL became the first programming language mandated by the DoD for use. This was likely a major factor in its survival, as the necessity of using the language kept its popularity up even though its early compilers were expensive and of poor quality. Calling it a government-subsidized programming language may be a little much, but having the DoD as a patron no doubt led to its widespread use and success, despite its lack of elegance and functions.

25.10 BACK TO THE BASICS

When modern-day technicians hear BASIC, they probably think of Visual BASIC, as in Microsoft's VB6, VBA, and VB.NET. But the original BASIC thrived in the 1970s and early 1980s. BASIC (Beginner's All-Purpose Symbolic Instruction Code) was designed by John G. Kemeny (1926–1992) and Thomas E. Kurtz (1928–) at Dartmouth College in 1964. The language was designed for use at computer terminals and was intended to be easy for nonscience students to learn, as science and engineering students made up only about 25% of the student population [1]. It achieved its goals; it was a very limited language and quite easy to learn. However, due to its simplicity and "friendliness," it was all too easy to write poor programs, and BASIC tended to perform quite poorly in the areas of readability and reliability.

Limitations and simplicity aside, perhaps the most important new aspect of BASIC was that it was the first widespread language to be used from terminals connected to a remote computer that ran all of the programs received from each terminal.

BASIC made a big comeback in the 1990s with the introduction of Visual BASIC (VB). The visual aspect provided a simple way to build graphical user interfaces (GUIs), and thus VB became widely used. As hinted at above, the most advanced current version is VB.NET, which departs significantly from classic VB due to its need to fit in with the .NET Framework and the rest of the .NET languages. One major difference is that VB.NET fully supports object-oriented programming. With the release of Microsoft's .NET Framework 4.0 in 2010, everything that can be done in VB.NET can be done in C#, and vice versa.

25.11 OVERVIEW: LOGICAL PROGRAMMING

As mentioned briefly in the description of declarative programming, logical programming languages are generally declarative (as opposed to imperative or procedural) in style. That is, the program doesn't specify how to reach a result, but rather describes the necessary characteristics of the result to be achieved and then logically infers an answer based on those attributes and on its existing tools or knowledge.

Predicate calculus notation is generally used in current logical programming languages, and provides a basic form of communication.

25.12 PROGRAMMING LOGIC: PROLOG

Prolog has a method for specifying predicate calculus propositions, and it implements a restricted form of resolution. The first Prolog system was developed by Alain Colmerauer (1941–) and Philip Roussel (1945–) in 1972 [5]. Prolog programs consist entirely of collections of statements, which make it easy to use to model an intelligent database of related information that can be queried. Prolog's statements can be either facts or rules. A Prolog program simulating an intelligent database would consist of a collection of both facts and rules, and could be queried with a question, a goal statement, which is presented to the program. The program then uses this goal statement along with its tools of inference and resolution to determine the truth of the statement. If it can prove based on the statements and rules at its disposal that the statement is true, it will do so. If not, it will conclude that the statement is false.

While simple and powerful, programs written in Prolog and other logical languages tend to lag in efficiency behind comparable imperative programs. Additionally, while logical programming is an effective approach for some kinds of database management systems and in some areas of AI, its usefulness is much more limited in other areas.

25.13 OVERVIEW: OBJECT-ORIENTED PROGRAMMING

Object-oriented programming (OOP) is a programming paradigm that uses objects to design and implement programs. An object is a collection of attributes and methods. Objects can be related to one another through concepts such as abstraction, virtualization, inheritance, and polymorphism. OOP is a common and widespread programming design, and many modern programming languages include support for OOP features at some level.

Objects control access to their attributes and methods, allowing as much or as little access as desired, with levels of access varying depending on the object or function that is attempting the access. Programs often contain many different types of objects, which generally represent real-world objects or concepts. One example would be a bank object representing an actual bank that contains a collection of account objects representing actual bank accounts.

At a minimum, an object-oriented language must provide support for abstract data types, inheritance, and dynamic binding of method calls to methods [1].

25.14 THE FIRST OBJECT-ORIENTED PROGRAMMING LANGUAGE: SMALLTALK

Smalltalk-80 (or simply Smalltalk) was the first language to fully support object-oriented programming (1980). All computing in Smalltalk is achieved by sending a message to an object to invoke one of its functions. The message replies themselves are objects that either return requested information or notify the sender that the requested action has been completed.

In Smalltalk, classes are abstractions of objects and can be instantiated as objects in the program. An object is always an instance of a class. This is a fundamental concept of object-oriented programming that propagated into many subsequent languages.

Unlike in common imperative and object-oriented hybrid languages such as C++ and Java, all values in Smalltalk are objects, even primitive values such as integers, characters, and Booleans. They are all instances of their corresponding classes, and all operations are invoked on them by sending messages. Because all values in Smalltalk are objects, classes are objects as well. Each class is an instance of the metaclass of that class, and each metaclass is an instance of the parent/root class metaclass.

In addition to pioneering the popular and widespread notion of object-oriented programming, windowed user interfaces (by far the dominant GUI design of the time) grew out of Smalltalk.

25.15 IMPERATIVE AND OBJECT ORIENTED: C++

Released in 1984, C++ improves on the imperative features of C and includes support for object-oriented programming. The classes in C++ are related to those in Smalltalk.

Improvements included virtual methods, operator and method overloading, and reference types. In version 2.0, support for multiple inheritance (classes can have more than one parent class), and abstract classes were added. In version 3.0, templates providing parameterized types were added, as was exception handling [1].

C++ quickly became a widespread language and remains widely used today. It is almost completely backward compatible with C and linkable to C programs. This feature certainly aided in the success of C++. It was also the only robust language available when object-oriented programming stepped into the spotlight, which made it a shoo-in for many large-scale commercial software projects.

25.16 OBJECT ORIENTED, HOLD THE IMPERATIVE: JAVA

Java was based on C++, and its designers sought to remedy some of C++'s drawbacks. What they removed, changed, and added led to a smaller and safer language with much of the flexibility and power of C++ still intact. Unlike C++, Java supports only object-oriented programming and does not support procedural programming.

One example of an unsafe feature of C++ is coercions. A coercion is an implicit type conversion. Both Java and C++ allow the coercion of smaller types into a larger type, for example, the coercion of an integer into a float. However, C++ also allows the coercion of a larger type into a smaller type, which can result in unintended data loss. Java does not allow this type of implicit conversion, reducing the risk that programmers will unintentionally lose data or precision.

Java was designed by Sun Microsystems to act as the programming language of choice for devices embedded in consumer electronics such as TVs and microwaves. Reliability was a chief concern, due to the very high cost that would be incurred if a mass recall of such devices was found to be necessary [1].

Object-oriented support was needed, but C++ was larger, riskier, and more complex than necessary. The new language designed for this purpose was simpler and more reliable. Ironically, none of the products for which Java was designed were ever marketed or sold. However, Java was found to be useful in web programming in the early 1990s when the Internet grew in popularity. Specifically, Java applets became very popular in the mid to late 1990s because they were simple and easy-to-use tools. The use of Java increased faster than that of any other programming language [1]. The most recent version is Java 7, introduced in 2011.

25.17 THE BEST OF BOTH WORLDS: C#

C# is based on C++ and Java, and also includes some concepts from Delphi (object-oriented Pascal) and Visual BASIC. C# is a multiparadigm programming language that includes imperative, declarative, functional, objected-oriented, and component-oriented programming features. C# was developed by Microsoft and later approved as a standard by European Computer Manufacturers Association (ECMA).

The ECMA standard lists these design goals for C# (ECMA International) [2]:

- C# is intended to be a simple, modern, general purpose, object-oriented programming language.

- The language, and implementations thereof, should provide support for software engineering principles such as strong type checking, array bounds checking, detection of attempts to use uninitialized variables, and automatic garbage collection. Software robustness, durability, and programmer productivity are important.

- The language is intended for use in developing software components suitable for deployment in distributed environments.

- Source code portability is very important, as is programmer portability, especially for those programmers already familiar with C and C++.

- Support for internationalization is very important.

- C# is intended to be suitable for writing applications for both hosted and embedded systems, ranging from the very large that use sophisticated operating systems to the very small having dedicated functions.

- Although C# applications are intended to be economical with regard to memory and processing power requirements, the language was not intended to compete directly on performance and size with C or assembly language.

As one of Microsoft's .NET languages, C# supports component-based software development in the .NET Framework. All .NET languages use the Common Type System (CTS) and are compiled into the same Intermediate Language (IL). A just-in-time compiler translates the IL into machine code before execution. All types in all of the .NET languages inherit from the shared class root System.Object. As mentioned earlier, with the release of C# 4.0 and VB.NET 4.0, anything that can be done in one language can be done in the other, which is evidence of Microsoft's effort to improve each of its .NET languages concurrently and cohesively.

In several areas, C# adds back in what Java stripped out of C++, but only after improving upon features and making them safer and more useful. C# also improves on features found in Java.

- Enums are back and are safer than those in C++. Enum values are no longer implicitly converted into integers, making them more type-safe.

- The struct is back and is actually useful, after having no real value in C++.

- C#'s switch statement is an improvement to the one found in C, C++, and Java.

- C# improves on C++'s unsafe function pointers by providing a new type called a delegate. Delegates are used for implementing event handlers and controlling thread execution.

- Methods in C# can receive a variable number of parameters of the same type.

- C# makes the conversion between the two distinct typing systems of C++ and Java partially implicit.

- C# features support for rectangular arrays.

- C# features a foreach statement, which is an iterator that can be used for collections of any type of object, including arrays.

- Properties replace public data members. Properties have built-in get and set methods that are implicitly called when references or assignments are made.

- C# has access to the resources and capabilities of the entire .NET Framework.

C# is truly a multipurpose programming language and has continued to rapidly evolve since its release in 2002. It is and looks to remain a versatile and widely used programming language.

REFERENCES

1. Robert Sebesta, *Concepts of Programming Languages* (10th ed.), Addison Wesley, New York, 2012.
2. *C# Language Specification* (4th ed.), ECMA International, 2006. http://www. ecma-international.org/publications/files/ECMA-ST/Ecma-334.pdf.
3. Wikipedia, Lisp (Programming Language), http://en.wikipedia.org/wiki/ Lisp_(programming_language).
4. Wikipedia, ALGOL 60, http://en.wikipedia.org/wiki/ALGOL_60.
5. Wikipedia, Prolog, http://en.wikipedia.org/wiki/Prolog.

CHAPTER 26

Algorithms and Computational Complexity

As described in Chapter 5, the word *algorithm* comes from the name of an Arabic mathematician, al-Khawarizmi, who wrote his famous book, *Rules of Restoration and Reduction*, in the ninth century. According to the book by Donald E. Knuth (1938–) [1], the original meaning of the word *algorithm* is "the process of doing arithmetic using Arabic numerals." Knuth also mentioned in the same book that the word *algorithm* did not appear in *Webster's New World Dictionary* as late as 1957. Let us look up the word in some old versions of English dictionaries. The *Concise Oxford Dictionary* in 1964 defines the word *algorithm* as Arabic decimal notation. The same dictionary in 1976 defines the word *algorithm* as the process or rules for calculation, almost the same definition used today by the *Concise Oxford Dictionary*.

Many university textbooks on algorithms often refer to Euclid's algorithm as an example of the first well-defined mathematical algorithm. It appears in the famous Greek book *Euclid's Elements* (c. 300 BC) (see Chapter 5). Euclid's algorithm is an effective and efficient procedure (a set of rules) for computing a two-argument function $gcd\ (m, n)$ determining the greatest common divisor of two integers m and n. We can informally say that an algorithm is an effective procedure for solving a problem. As described in Chapter 16, Alan Turing gave a formal definition of an algorithm in 1936 in terms of abstract models of computers, called Turing machines. Alonzo Church, Stephen C. Kleene, Emil L. Post, and others also gave formal definitions of algorithms in different forms and computational models, although all these definitions were eventually shown to be equivalent.

A computer is a physical device. In order to carry out computation, we must supply a procedure or a method of how the computation is to proceed. Such a procedure should then be written in a programming language so that the computer can understand how it should process data in its registers or memory at each step. Chapter 27 discusses the specifics of algorithmic design.

An automobile or an airplane needs fuel to move. For example, gasoline is one possible resource for an engine to operate. Analogously, for the engines of computation, we also need to consider resources that are needed to carry out their tasks, even though they may only be abstract models that carry out the computation. In the area of theoretical computer science, Turing machines are usually used as abstract models of computing devices. In computations, the two most common measure types of resources are *time* and *space* requirements. These are often used to evaluate the efficiency of an algorithm or the difficulty of a problem. Time and space complexity are indications of how much computing time (the number of steps) and memory (space) are required, respectively, to carry out the computation of a given algorithm in order to solve a problem.

The *complexity* (time or space) *of an algorithm* is represented by a resource binding function whose parameter is the size of a problem instance, say n. This function depicts the amount of resource (time or space) needed to solve the problem instance by the algorithm. The *complexity of a problem* is represented by the complexity of the best algorithm for solving the problem. Alternatively, one can discuss the *lower bounds* of a problem, which indicate the least amount of resources (time or space) needed by *any* algorithm (already existing or not) needed to solve a given problem. If the complexity (e.g., the running time) of an algorithm is the same as the problem's complexity, then the algorithm is said to be *optimal* with respect to the particular resource complexity measure. Strangely enough, it can be shown that there do exist problems for which no optimal algorithms exist.

Since the mid-1950s, computational complexity has become a very active research topic, and it has developed into one of the most fundamental areas in theoretical computer science. Let us explain the meaning of computational complexity using a simple example of the problem of addition of two numbers (i.e., $f(x, y) = x + y$). Assume we specify the values for x and y using the decimal notation. Then the *size* of a problem instance in this case is the number of digits needed to represent x and y. For the binary, hexadecimal, or any other reasonable notation, the size of a problem instance is similarly defined. Consider the conventional addition algorithm taught in elementary schools: its time complexity is clearly proportional to the argument size, that is, the total number of digits used to represent x and y.

In general, the actual computing time of an algorithm for solving a problem depends on the performance of the computing device or the computing model. The addition of two numbers is usually performed by aligning them and adding their least significant digits, moving to the second least significant digits (with a carry, if applicable), adding them, and so on, until the most significant digits have been added. If the computing time at each stage is considered to take a *unit time* (or a step), then the computing time needed for adding two numbers is at most the number of digits of the larger one (assuming both are positive).

In the addition algorithm mentioned above, the time complexity is proportional to the input size, and it is denoted by $O(n)$, where O notation is pronounced "big oh" or "at most order of." Suppose that a new computing device can concurrently process the addition of some fixed number of consecutive digits of the two numbers in a single unit of time. This computing device may seem to be more powerful, i.e., quicker, than the original one. However, the time complexity of the addition of two numbers by this more powerful device is still $O(n)$, since for any positive integer k (independent of n), $O(n/k) = O(n)$. From this observation, the reader should be convinced that the big oh notation for complexity classes is independent of the computing device used. For example, if we build a new computer that is 1000 times faster than the previous one, the big oh running times for any particular algorithm on these two computers will, in fact, be the same. This makes the big oh notation very convenient.

Next, let us consider the time complexity of integer multiplication, $f(x, y) = x \times y$. Assume that both the multiplication and addition of two single-digit integers can be performed in a single time unit. Let n be the number of digits of the larger number among x and y. It should be obvious that the conventional multiplication algorithm that is taught in elementary schools can be carried out in $O(n^2)$ steps.

One could argue that since we ourselves have designed/constructed the algorithms, it should be relatively easy to determine their running times. That, unfortunately, is not always the case—one could easily write computer programs whose running times themselves are very difficult, if not impossible, to determine.

To make things even worse, in general, the evaluation of the complexity of a problem, instead of an individual algorithm, is much more difficult. The time complexity of the integer addition is obviously $O(n)$ since every digit of x and y, as well as the digits of the resulting sum, should be processed at least once to obtain the correct answer. However, the complexity of integer multiplication cannot be determined using such a simple argument. In 1964, Russian mathematicians A. Karatsuba (1937–) and Yu Ofman discovered a more efficient algorithm for integer multiplication. Their algorithm uses the *divide-and-conquer* approach, and its time complexity is $O(n^{1.59})$, where 1.59 is from an approximation of $\log_2 3$. The divide-and-conquer technique divides a big-sized problem instance into smaller-sized problem instances, and the solution to the original problem instance is obtained from the solutions to the smaller-sized problem instances. This method is used recursively until we obtain sufficiently small problem instances that can be easily handled by some simple method. The above-mentioned Karatsuba–Ofman algorithm for integer multiplication is not even the best one available. The Schönhage–Strassen algorithm for integer multiplication is asymptotically fast for large integers. It was developed by Arnold Schönhage (1934–) and Volker Strassen (1936–) in 1971. The algorithm uses the *fast Fourier transforms*, and its running time is $O(n \log n \log \log n)$, as opposed to $O(n^2)$, for the naïve integer multiplication algorithm. From the function graphs shown in Figure 26.1, we can easily understand that the Karatsuba–Ofman algorithm is much faster than the

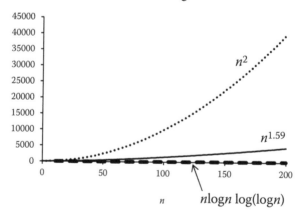

FIGURE 26.1   Comparison among functions $n \log n \log \log n$, $n^{1.59}$, and n^2.

conventional method, and the Schönhage–Strassen algorithm is much faster than the Karatsuba–Ofman algorithm [9].

Computational complexity is at the core of theoretical foundations of computer science. The field analyzes the complexity of problems and their algorithmic solutions. The major questions in this field are concerned with what can be achieved within limited computational resources. The most influential work at the beginning was the formal definition of computability by Alan Turing in 1936. The beginning of systematic study in computational complexity started around the 1960s. Since then, a number of interesting results have been obtained. Juris Hartmanis (1928–) and Richard Stearns (1936–) introduced time complexity classes, and proved the *time hierarchy theorem* in 1965 [5]. The theorem ensures the existence of certain difficult/hard problems, which cannot be solved in a given amount of time. This means that for every time-bounded complexity class, there is a strictly larger class containing problems that are not members of the smaller class. A similar argument may be presented for space-bounded complexity classes. Hartmanis and Stearns received the Turing Award in 1993 for their contributions in the field of computational complexity theory.

Manuel Blum (1938–) developed an axiomatic complexity theory, called the Blum axioms or *Blum complexity axioms*, which specifies desirable properties of complexity measures on the set of computable functions in 1967 [6]. He proved that for a complexity measure satisfying Blum axioms, some fundamental properties must always hold true. Blum received the Turing Award in 1995 for his contributions to the foundations of computational complexity theory and its application to cryptography and program checking.

The big oh notation and a recursive function $f(n)$ can be used to define the computational complexity class $O(f(n))$. Computational complexity classes defined in this way seem to be coarse since $O(f(n))$ and $(O(g(n))$ define the same complexity class if there exist positive constants k, c_1, and c_2 such that for any $n > k$, $c_1 g(n) < f(n) < c_2 g(n)$. However, this coarseness is rather convenient for the study in this field. For example, if computing model A is twice as fast as computing model B, and if the computing time by A is $f(n)$, then the computing time by B is $2f(n)$ and $O(2 f(n)) = O(f(n))$. This means that concerning computational complexity classes, we do not care what computing models are used, provided that the differences among their computing abilities are within a constant factor.

A problem X is said to be *complete* for complexity class C, or *C-complete*, if it is the hardest problem within C; i.e., every other problem in C can be shown (via a complexity-preserving transformation) to be no more difficult than X. Observe that more than one problem may be complete for a given complexity class. A problem Y is said to be *hard* for a class if every problem within the class is no more difficult than Y. Note that Y does not have to be a member of the given class.

Over the years, the notion of *efficient computation* or an *efficient algorithm* has been associated with two distinct definitions. The first one describes an algorithm as efficient if it can be shown that during its computation no unnecessary work is being performed. This, however, does not mean that a given efficient algorithm must always be optimal with respect to some single complexity measure (e.g., time). In certain applications, the computational resources of both time and space are constrained, and in order to meet both of them simultaneously, an algorithm may have to perform additional steps or use extra space in order to perform its computation successfully. For many problems, we can show that there exists a *time-space trade-off*, meaning, indirectly, that optimality criteria for both resources cannot be achieved at the same time.

The second definition of an efficient algorithm, which gained wide acceptance during the 1970s, is that it's an algorithm whose running time is bounded by some polynomial function in n, the problem size. More formally, an algorithm is of polynomial time complexity if it belongs to computational complexity class $O(n^c)$ for some constant $c > 0$, which is independent of n. If there exists a polynomial time algorithm for solving a given problem X, then X belongs to the class of polynomial time problems, which is denoted by **P**. A problem is said to be *infeasible*, or *provably intractable*, if it does not belong to **P**. Note that a problem may be intractable, but still solvable at the same time. If a problem does not belong to **P**, we consider that it cannot be solved in the practical sense. This, of course, does not mean that for such problems we cannot try to use approximation or heuristic algorithms, which in certain instances may give us a satisfactory or even an optimal answer to our problem. The computational class **P** can also be defined in terms of Turing machines. A problem belongs to **P** if and only if there exists a Turing machine that solves the problem in polynomial running time.

Until now, the computations carried out by our Turing machines (or other computational devices) have been *deterministic*—meaning that the steps taken always follow the same predetermined computational path. A *nondeterministic* Turing machine is an enhanced Turing machine with an additional power of making choices. For a given state and a tape symbol scanned by the tape head, the nondeterministic Turing machine has some finite number of possible choices, or moves, for the next step. A problem is said to be of nondeterministic polynomial time complexity if and only if it can be computed by a nondeterministic Turing machine that runs in polynomial time. The class of nondeterministic polynomial problems is denoted by **NP**, and it obviously contains the class **P**. Also, it is not difficult to show that any problem in **NP** can be solved by a deterministic Turing machine that runs in exponential time. However, it is yet to be determined whether **P** = **NP**. This question was originally formulated by Stephen Arthur Cook (1939–) in 1971 [7], and it is the most famous open problem in theoretical computer science and mathematics. The **P** = **NP** question can be rephrased as asking whether there exists a deterministic Turing machine with polynomial running time for any one of the **NP**-complete problems.

The class **NP** can be thought of as those problems for which there exist algorithms (or Turing machines) that can verify in deterministic polynomial time if a *guess* or a *certificate* (that is provided as part of the input describing the problem) confirms the answer. One might name this class as that of *conscientious cheaters*. Consider the problem of inverting a matrix. During a linear algebra exam, one student (who always obtains the correct answer) may use a standard technique, such as Gaussian elimination, to obtain the inverse; another student may cheat and simply copy the first student's answer. However, a conscientious cheater would (after copying the first student's answer) actually check and verify whether the answer is correct.

The question of whether it is always more difficult to compute an answer, rather than simply check the correctness of a proposed answer, was first posed by German mathematician Johann Carl Friedrich Gauss (1777–1855). While it is obvious that the computation of an answer to any problem *is at least as difficult* as confirming its correctness, it has not been shown that it must always be necessarily *more* difficult.

NP-complete problems are the most difficult problems within the class **NP**; moreover, all of them have the property that if we could construct an efficient (deterministic polynomial time) algorithm for one of them, then all **NP**-complete problems would have efficient algorithms and **P** would indeed be equal to **NP**. On the other hand, if one **NP**-complete problem can be shown to be outside of **P**, then none of the **NP**-complete problems would have efficient algorithms.

The first **NP**-complete problem was identified by Cook in 1971, and it dealt with the satisfiability of Boolean expressions. A set of logical propositions is said to be *satisfiable* if there exists at least one assignment to a list of Boolean variables so that every clause in the set evaluates to "true" under that assignment. The *satisfiability problem* (**SAT**) is to decide whether a given set of Boolean expressions is satisfiable. In 1971 Cook showed that **SAT** is the hardest problem in the class **NP**; i.e., it is **NP**-complete. If **SAT** were to be a member of **P**, then **P** = **NP** (which is very unlikely, although still a possibility).

Any **NP**-complete problem must:

1. Belong to the class **NP**

2. Be at least as difficult as some known **NP**-complete problem

If a problem *X* can be reduced to another problem *Y* in polynomial time, and if *X* is **NP**-complete and *Y* belongs to **NP**, then *Y* is also **NP**-complete. The technique, or transformation, used here is called the *polynomial time reduction*. After Cook's initial result, many problems were added to the list of **NP**-complete problems via polynomial time reductions. It should be noted that Richard Karp (1935–) and Leonid Levin (1948–) independently made equally significant contributions during the early developments of the theory of **NP**-completeness in the early 1970s.

The collection of **NP**-complete problems is ever growing, and it contains tens of thousands of problems. After **SAT**, the *traveling salesman*, *independent sets*, *Hamiltonian circuit*, and *knapsack* are some of the best-known **NP**-complete problems. When we show a problem is **NP**-complete, it is a good indication that there is very little chance to find an efficient algorithm for it. For such a problem, a heuristic method or an approximation algorithm may be recommended. A book by Michael R. Garey and David S. Johnson (1979) contains an excellent introduction to the theory of **NP**-completeness and a listing of several hundred **NP**-complete and other related problems [8].

The reader may wonder if there could exist problems in **NP** that are outside of **P**, but which are not **NP**-complete, i.e., difficult, but not the most difficult. The status of this question is open as well. In fact, there are not too many problems in **NP** whose status is unknown (i.e., are they in **P** or are they **NP**-complete?). One of the most famous of these is the *graph isomorphism problem* (GIP)—given descriptions of two graphs, do they have the same structure, or connectivity, relationship? For many restricted cases, such as planar, degree-bounded graphs, etc., efficient/polynomial algorithms have been found. However, the general GIP is still not known to be in **P**, nor has it been shown to be **NP**-complete.

Just as intriguing is the question whether there exist problems that provably require more than polynomial time (deterministic or not) to be solved. The answer is yes; however, the only known such problems also require an exponential amount of space (memory), making them very unattractive from the practical perspective. Determination of "forced wins" in strategic games such as chess or Go can be shown to require an exponential amount of space, and hence time, to compute.

Stephen A. Cook received the Turing Award in 1982 for his formalism of **NP**-completeness and discovery of the first **NP**-complete problem, **SAT**. Richard Karp received the Turing Award in 1985 for his introduction of the now standard methods for proving problems to be **NP**-complete, which has led to the identification of many theoretical and practical problems as being computationally hard.

More material on computational complexity can be found, for example, in the books by Michael Sipser [2], Sanjeev Arora (1968–) and Boaz Barak [3], and Oded Goldreich (1957–) [4].

REFERENCES

1. D. E. Knuth, *The Art of Computer Programming: Fundamental Algorithms* (vol. 1, 2nd ed.), Addison-Wesley, Reading, MA, 1973.
2. M. Sipser, *Introduction to the Theory of Computation* (2nd ed.), Course Technology, Boston, MA, 2006.
3. S. Arora and B. Barak, *Computational Complexity: A Modern Approach*, Cambridge University Press, New York, 2009.
4. O. Goldreich, *Computational Complexity: A Conceptual Perspective*, Cambridge University Press, New York, 2008.
5. J. Hartmanis and R. E. Stearns, On the Computational Complexity of Algorithms, *Transactions of the AMS*, 117, 285–305, 1965.
6. M. Blum, A Machine-Independent Theory of the Complexity of Recursive Functions, *Journal of ACM*, 14, 322–336, 1967.
7. S. A. Cook, The Complexity of Theorem Proving Procedures, in *Proceedings of 3rd Annual ACM Symposium on the Theory of Computing*, 1971, pp. 151–158.
8. M. R. Garey and D. S. Johnson, *Computers and Intractability: A Guide to the Theory of NP-Completeness*, W. H. Freeman, New York, 1979.
9. A. V. Aho, J. E. Hopcroft, and J. D. Ullman, *The Design and Analysis of Computer Algorithms*, Addison-Wesley, Reading, MA, 1974.

CHAPTER 27

The Design of Computer Algorithms

As described in Chapter 18, computers were traditionally built to calculate the solution to numerical problems such as the determination of the roots of an algebraic equation, or to process numerical data, e.g., the statistical calculation of national consensus data. However, for the last 40 years or more, a great deal of important work has also been done by computers for nonnumerical problems, such as sorting, searching, word or language processing, and solving combinatorial problems in software development, data communication, and the simulation of various biological, chemical, and physical processes. The design and analysis of computer algorithms have been developed in the context of such applications.

27.1 SORTING AND SEARCHING

Sorting a sequence of numeric or alphanumeric items means rearranging them in some specific order. Since a significant portion of data processing involves sorting, efficient sorting algorithms are of considerable importance for many practical problems [3, 7]. Sorting is one of the first problems for which algorithms were formally analyzed and evaluated. It is considered to be a fundamental problem in the area of *algorithmics*, mainly because sorting often plays an essential part of many algorithms. For example, sorting significantly facilitates a search for elements in some set or a list. A dictionary is a typical example of a sorted list. Words in a dictionary are usually arranged in an alphabetical order, which allows us to easily locate the word that we are trying to find in the dictionary.

Some sorting methods were in use long before the invention of computers. For example, a card player sorts the cards of his hand so that each card is visible and individually accessible. Suppose that *A*, 2, 3, 4, 5, 6, 7, 8, 9, 10, *J*, *Q*, *K* are the order of the cards. The following method has been commonly used among card players. The player scans the cards of his hand left to right. He chooses the smallest card, and puts it at the leftmost position. Then he scans the rest of the cards, and chooses the smallest one among them. This card is placed at the second leftmost position. In this way, he arranges the smallest card, the second smallest card, the third smallest card, and so on. Eventually he obtains the sorted cards in his hand. This sorting method is called *straight selection sorting* or *sorting by selection*. We will soon find out that it is not the fastest way to sort the elements of a given set.

The *efficiency* of a sorting algorithm is often measured by counting the number of necessary key comparisons in order for the list to be sorted. Usually, this is expressed as a function of the number *n* of items in the list. Sorting by straight selection requires $n - 1$ key comparisons to choose the smallest item among *n* items. Then, $n - 2$ key comparisons are needed to choose the second smallest item from the remaining $n - 1$ items, and so on. Consequently, sorting *n* items by the straight selection method requires $(n - 1) + (n - 2) + \ldots + 1 = n(n - 1)/2$ comparisons. Using the notation for complexity classes given in Chapter 26, the efficiency of sorting by straight selection is $O(n^2)$ time. Simple and obvious sorting methods such as sorting by selection, sorting by insertion, and sorting by straight exchange (also called the *bubblesort*) all require $O(n^2)$ key comparisons [4, 5].

It can be easily proved that any sorting algorithm based on a comparison operation between a pair of items requires at least $O(n \log n)$ comparisons; if it uses less, then it is provably incorrect. *Quicksort* is a well-known sorting algorithm that, on the average, needs $O(n \log n)$ comparisons to sort *n* items. In the worst case, however, it needs $O(n^2)$ comparisons, though if it is properly implemented, such a case is very rare. In practice, quicksort is considerably faster on the average than other $O(n \log n)$ sorting algorithms such as *heapsort* and *mergesort* [1, 4, 8].

Quicksort was invented in 1960 by a British computer scientist, Charles Antony Richard Hoare (1934–), when he was in the Soviet Union as a visiting student at Moscow State University. After Hoare left the Soviet Union, he began working with British computer manufacturer Elliot Brothers, where he implemented ALGOL 60, one of the first high-level programming languages (See Chapter 25). In 1968, he moved to Queen's University of Belfast in 1968, and in 1970 moved to Oxford University as a professor of computing. Hoare received the Turing Award in 1980 for his contributions to the definition and design of programming languages [10, 11].

As a first step, quicksort chooses one of the items in the sequence of items as the pivot element. If the pivot is chosen randomly, the quicksort is called the *randomized quicksort*. The sequence is then partitioned on either side of the pivot so that the items that are greater than the pivot are placed on its right section, whereas all the other items are placed on its left section. Next, the two sections of the sequence on either side of the pivot element are sorted independently by recursive calls of the algorithm. The final result is a completely sorted sequence. We should be careful to choose a pivot so that the pivot should partition the list into two sections with balanced sizes. An algorithm (or a procedure) that calls itself is said to be *recursive*. ALGOL 60 was the first programming language in which recursive calls were implemented. Hoare noticed that quicksort can be easily implemented in ALGOL 60 by using recursive calls.

Storage and retrieval of data are the fundamental tasks in data processing. We are concerned with collecting data efficiently into the computer memory, and are often asked to retrieve the necessary data from computer memory as quickly as possible. Searching is a process to find or decide whether specific data exist in the computer memory. It is a very time-consuming task concerning many software applications, and good search methods are important to improve software performance in general.

It is often possible to arrange the data or data structure so that we can quickly know where a necessary item is located. If the data are neither structured nor sorted, we must sequentially search all of the data to find the item we need. In such a case, we need $O(n)$ time to search for an element in O set or list of n items. If the data are sorted, we can apply a *binary search* method (e.g., as in finding a word in a dictionary). In this case, we only need $O(\log n)$ time to find the item we need, which is a significant saving of time.

Search methods can be either *static* or *dynamic*. Static means that the contents of the data files are essentially unchanged. For this type of data file, it is important to minimize the search time without regard for the time required to set them up. On the other hand, dynamic means that the contents of the data files vary frequently by the data insertions and data deletions. For this type of data file, suitable set manipulation algorithms are required to keep good structures of the data files.

Rich material about sorting and searching, including their history, can be found in Knuth's *The Art of Computer Programming: Sorting and*

Searching [4]. Knuth announced that he would publish a series of seven volumes entitled *The Art of Computer Programming*. Volume 1 of the book series was first published in 1968 [2]. After the publication of Volume 3 in 1973, he developed an electronic typesetting system, creating the now widely used Tex and METAFONT tools. Knuth received the Turing Award in 1974 for his research contributions to the analysis of algorithms, the design of programming languages, and computer programming through the series of his books.

27.2 DATA STRUCTURES

Efficient ways to store and access large amounts of data are an important part of many computer applications. *Data structures* are to organize the data in a computer so that they can be accessed and manipulated efficiently.

Often a problem can be formulated in terms of abstract objects such as sets, lists, graphs, and so on. Algorithms for a problem usually contain fundamental operations on these objects. Typical examples of these operations are membership determination, insertion of a new element into a set, deletion of a member from a set, union of sets, and determination if a set contains a particular element. In order to implement these operations efficiently, various data structures have been devised. If needed, a higher-level *compound data structure* may often be built from these fundamental data structures.

Arrays and *lists* (also called *linked lists*) are the most fundamental data structures. An array is a linear structure, all of whose elements/components are of uniform size and type. It is also called a random access structure since all components are accessible in one time step. A linked list is a collection of *nodes* (or *cells*) arranged linearly in a certain order. The length of a linked list is not fixed. The linked list data structure must allow us to efficiently determine (via a pointer) the first and last nodes in the list, and which nodes are the predecessor and the successor (if they exist) of any given node. Its access mechanism is said to be sequential. Such a structure is often represented graphically by boxes and arrows, as shown in Figure 27.1. The information attached to a node is shown in the corresponding box, and the arrows show the relationship between a node and its successor.

The head of a list

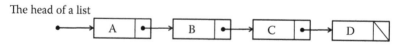

FIGURE 27.1   An example of a list.

A general method for constructing structured types of data is to join elements into a compound cell. Such a cell is frequently called a record or an object, and it is usually consists of various components [5]. For example, consider a table of records for students in a class. A record for a student consists of various components, e.g., student's name, age, birthplace, and academic grade for each course. Suppose that this table is sorted in alphabetical order by students' last names. If the number of students in the class is fixed, then an array of records would be a convenient data structure to use. However, in the case where the class size is dynamic, an array of records is inconvenient and inefficient as well. When a new student joins the class, the record of the new student should be inserted at an appropriate (alphabetical) position. In order to do so, all the records of the students after the new student, in alphabetical order, should be shifted by one position so that the space for the record of the new student is made available. In the case where a student drops the class, the removal of the record for the student creates an empty space in the array. We must then move up the records for the students one position in the array after the record for that student, which could be quite time-consuming.

The data manipulated by algorithms can frequently grow, shrink, or change over time. For such dynamic data, a linked list is more convenient and efficient, because the insertion and deletion operations can be implemented by simply modifying the corresponding pointer that is used for the connection to the successor or the predecessor. For example, in order to remove an element x from a linked list, the pointer to x is modified to delete x out of the linked list in a way that it points to the element that was originally pointed to by x. In order to insert a new element y after element x in a linked list, the pointer from x is modified to point to y, and then the pointer of y is created to point to the element that was originally pointed to by x.

Many algorithms need the ability to insert elements into, delete elements from, and test memberships in a set. However, other algorithms may require more complicated operations to be efficiently implemented. For such algorithms, various advanced data structures have been devised. For example, *priority queues*, *heaps*, *hashing schemes*, *B-trees*, and *Fibonacci heaps* are some advanced data structures that allow us to solve complex problems more efficiently [8].

A Programming Language (APL) is an iterative array-oriented programming language invented in 1957 by Kenneth E. Iverson (1920–2004). Operations on structured arrays can be written in APL in a straightforward way. Iverson worked on programs that could evaluate large matrices on the Harvard Mark IV computer. In 1960, while at IBM, he developed his ideas into a programming language for the IBM/360. In 1979, Kenneth E. Iverson received the Turing Award for his work on APL.

Linked lists were developed in 1955–1956 by Allen Newman, Cliff Show, and Herbert Simon as the primary data structure for their data processing programs and early artificial intelligence (AI) programs [12]. In 1958, Victor H. Yngve (1920–) used linked lists at MIT as data structures in the computer programs for processing linguistic problems. LISP is a programming language invented by John McCarthy in 1958, while he was at MIT. It stands for List Processing. It is a suitable programming language for processing linked lists. It has numerous applications in AI and is a very powerful and expressive language, especially for nonprocedural applications, such as *functional programming* (see also Chapters 24 and 25). A Swiss computer scientist, Nicklaus Wirth, wrote an outstanding textbook, *Algorithms + Data Structures = Programs*, in 1976 [5]. In his book, he showed how to describe algorithms and data structures in PASCAL. That is, he showed that designing a nice program means designing a nice algorithm and a nice data structure. In 1984, he received the Turing Award for developing a series of innovative programming languages.

27.3 GRAPH ALGORITHMS

One major feature of computer science is its discrete flavor, which means that data processing and computing consist of a series of discrete operations on discrete and finite data. Discrete mathematics has developed to formalize basic combinatorial structures. One of the most fundamental ways to express a discrete structure is as a graph.

The 18th-century East Prussian town of Königsberg (now Kaliningrad, Russia) lay on the bank of the Pregal River and two islands connected by seven bridges, as shown in Figure 27.2. The people of Königsberg had

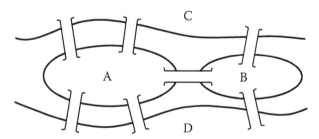

FIGURE 27.2   Königsberg's bridges.

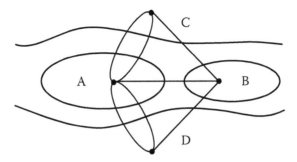

FIGURE 27.3   The graph representing Königsberg's bridges.

a question about walking routes crossing these bridges. The question
was whether it was possible to start walking from anywhere in town and
return to the starting place while crossing all seven bridges exactly once.

In 1736, the people of Königsberg wrote to a well-known Swiss mathematician, Leonhard Euler, about this question. Euler solved the problem
by proving that such a walk was impossible by formalizing it as a graph
problem. He replaced the islands and riverbanks by vertices (nodes) and
the bridges by edges. Then a *graph* (*multigraph*) was obtained as shown in
Figure 27.3. This graph is equivalent, for the purpose of the problem, to the
map in Figure 27.2.

Graph $G = (V, E)$ is a simple way of representing pairwise relations among
a set of objects (nodes, vertices). It consists of a set of vertices V and a set of
edges E. Each of the edges corresponds to a pair of distinct vertices. Thus,
we represent an edge in E as a pair of vertices, joining the two vertices.
If each edge in E is an ordered pair of vertices, then G is called a directed
graph (or digraph). The graph given in Figure 27.3 has multiple edges, since
there are two edges connecting vertices A and C, and two edges connecting vertices A and D. This type of graph is usually called a multigraph. The
degree of a vertex in a graph (multigraph) G is the number of edges in G that
are incident to the vertex.

A *path* in a graph or a multigraph G consists of an alternating sequence
of vertices and edges of the following form:

$$v_0, e_1, v_1, e_2, v_2, \ldots, e_{n-1}, v_{n-1}, e_n, v_n,$$

where each edge e_i contains the vertices v_{i-1} and v_i. The number of edges in
a path is called the *length* of the path. A path is called *simple* if each vertex
appears in the path at most once. A *circuit* is a path whose first and last
vertices are the same.

The question about walking the bridges of Königsberg is considered to have been the first problem in graph theory, whose creation is credited to Euler. It is equivalent to the graph-theoretical problem of whether it is possible to find a circuit in the graph that contains each edge exactly once. Euler solved this problem by proving that such a traversal is possible if and only if the graph is connected and all of its nodes have even degrees. Königsberg's bridge problem showed, in a sense, that a realistic problem may be modeled as a graph with some properties. In fact, many fundamental problems can be represented by graphs and then solved using appropriate graph algorithms.

Consider a problem for finding the shortest (least expensive) routes from the head office to each of its branch offices. The problem can be modeled as the *single-source shortest-path* problem in a weighted and directed graph, where each edge of the graph has a weight (i.e., the length or cost). The head office corresponds to the source vertex, and the length of the road directly connecting any two offices is represented by the corresponding weighted edge. An elegant algorithm to solve this problem was proposed in 1959 by Edsger Wybe Dijkstra (1930–2002) [14]. His algorithm for the problem is an example of a typical *greedy algorithm*. A greedy algorithm makes the choice that appears to be the best one at each step. It makes a locally optimal choice with the expectation that this choice will eventually lead to a global optimal solution. The time complexity of *Dijkstra's algorithm* for the shortest paths from the single source is $O(|V|^2 + |E|) = O(|V|^2)$, where $|V|$ is the number of vertices and $|E|$ is the number of edges. For the weighted and directed graph in Figure 27.4, the shortest paths from vertex 1 to every other vertex are $1 \to 3 \to 2$, $1 \to 3$, $1 \to 5 \to 4$, and $1 \to 5$.

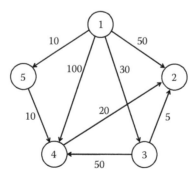

FIGURE 27.4 A weighted and directed graph.

It should also be mentioned that Dijkstra was well known for his criticisms of the "Goto" statement in computer programming, writing his opinions in various articles (e.g., "Go To Statements Considered Harmful," 1968). He was a leading person of a programming methodology called *structured programming*. It allowed for more understandable programs, which were more easily modifiable, and also made it easier to prove their correctness. Dijkstra received the Turing Award for his contributions to algorithms, programming languages, and programming methodology in 1972.

Michael L. Fredman and Robert E. Tarjan (1948–) improved Dijkstra's algorithm by using the Fibonacci heap data structure [13]. The time complexity of the improved version by Fredman and Tarjan for the shortest-path problem from the single source is $O(|V| \log |V| + |E|)$. Tarjan received the Turing Award for his contributions to the design and analysis of algorithms and data structures in 1986.

Dijkstra's algorithm may not work properly if negative edge weights are allowed in the graph. The *Bellman–Ford algorithm* solves the single-source shortest-path problem in the general case in which edge weights may be negative. The algorithm is based on algorithms proposed in 1958 by an American applied mathematician, Richard Bellman (1920–1984), and its refined version was proposed in 1962 by an American mathematician, Lester R. Ford (1886–1967). The time complexity of the Bellman–Ford algorithm is $O(|V||E|)$.

The problem of finding shortest paths between all pairs of vertices in a directed and weighted graph is also interesting. Of course, we can solve this problem by solving the single-source shortest-path problem for each vertex in the graph. This approach, however, is not recommended, because it is not the most efficient way to proceed. *Floyd's algorithm* for the all-pairs shortest paths was proposed in 1962 by Robert Floyd (1936–2001); it uses a dynamic programming approach based on matrix multiplication. The algorithm uses an adjacency matrix representation of a graph. The input is an $n \times n$ matrix W representing the edge weights of an n-vertex directed graph $G = (V, E)$, where $W = (w_{ij})$ is the weight of the edge from vertex i to vertex j.

Dynamic programming is a method of solving problems by breaking them down into overlapping subproblems. This method was originally proposed and developed by Bellman in the 1940s and refined by the early 1950s. The overlapping subproblems are of sizes that are smaller than the original problem. Each subproblem is then recursively broken down into yet smaller subproblems. For example, suppose that we wish to find the shortest path from A to B, and that possible intermediate vertices are M_1, M_2, and M_3. Then, we choose the best intermediate vertex from M_1, M_2, and M_3 by deciding the smallest value among dist(A, M_1) + dist(M_1, B), dist(A, M_2) + dist(M_2, B), and dist(A, M_3) + dist(M_3, B), where $dist(X, Y)$ means the shortest distance from vertex X to vertex Y. Floyd received the Turing Award in 1978 for his contributions to methodologies for the design of efficient and reliable software [6, 9].

The algorithms for the shortest-path problems are just a few examples of a large family of graph algorithms. Graphs can elegantly model a variety of discrete and optimization problems. As another example, we describe the problem of scheduling classes in a university or meetings of government committees as a graph-theoretical problem. The vertices correspond to the classes (the meetings), and two classes (two meetings) are connected by an edge if there is a student (or a government member) who wishes to attend both classes (meetings). If classes (meetings) are directly connected by an edge, then they should be held in different time periods or on different days. The problem is to schedule the classes (meetings) in a way that the conflicts are minimized. This problem can be modeled as a graph coloring problem, as shown in Example 27.1. It should be noted that most scheduling problems have been shown to be **NP**-complete and are therefore extremely difficult to solve.

Example 27.1

In a local government, there is an education committee (Ed), an environment committee (En), a finance committee (F), a housing committee (H), a security committee (S), a transportation committee (T), and a welfare committee (W). We wish to schedule the committee meetings so that the meetings of any two committees should be held on different days if there is a member belonging to both of them, and that the number of days when any meeting is held should be minimized. Each meeting corresponds to a vertex of the graph, and two vertices are connected by an edge if there is a member who belongs to both committees. Alternatively, we can describe this as a problem of finding the minimum number of colors needed to color each vertex of a graph so that any adjacent vertices should be colored differently. As shown in Figure 27.5, the minimum number of colors needed for this example is two. That is, it is possible that 2 days are sufficient to hold all of the meetings satisfying the condition.

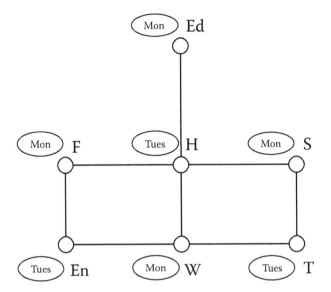

FIGURE 27.5   A graph for a meeting schedule of a committee.

27.4 RANDOMIZED ALGORITHMS

Usually computer algorithms behave predictably. In other words, most computer algorithms are deterministic. Given a particular input, a deterministic algorithm will produce the same output, and the underlying machine will always work through the same sequence of computing states. Consequently, for a given input, a deterministic algorithm requires the same execution time. Except for randomized quicksort, all the algorithms in the previous sections were deterministic.

As described in Section 27.1, randomized quicksort, devised by Hoare in 1960, contains random choice operations. Consider sorting a set by randomized quicksort. We first choose a random element y from S, and then partition $S - \{y\}$ into two sets S_1 and S_2, where S_1 consists of those elements of S smaller than y, and S_2 consists of remaining elements. S_1 and S_2 are then recursively sorted in the same way, eventually resulting in a sorted sequence of the elements of S. An algorithm containing random choice operations is called a *randomized* or *probabilistic* algorithm. The fundamental characteristic of randomized algorithms is that they may react differently if they are applied twice to the same input instance. Given the same input to the randomized quicksort, the computational time may vary different at different runs/executions, but the result obtained will be the same. However, sometimes, even the results may vary from one execution to the next of randomized algorithms, depending on which computational steps were taken.

Randomness was first used in algorithms for the approximate solution of numerical problems. It can be traced back to the 19th century. For example, it was used in the experimental determination of π around the year 1870. Stanislaw Marcin Ulam (1909–1984), who was an American mathematician of Polish-Jewish origin, used randomized computations in atomic physics research during World War II in Los Alamos, New Mexico. He had coined the phrase "Monte Carlo algorithms," which is still in use today. In 1976, a symposium entitled "Algorithms and Complexity: New Directions and Recent Results" was held at Carnegie Mellon University. At the symposium, Richard Karp (1935–) and Michael Oser Rabin (1935–) gave lectures on various topics. Karp presented an outline of randomized algorithms, and Rabin showed fundamental methodologies of how to design efficient randomized algorithms. This symposium has had a great influence on the development of subsequent research in the area of randomized algorithms. Michael O. Rabin received the Turing Award in 1976 for his 1959 work with Dana Scott (1932–), "Finite Automata and Their Decision Problems" and his many other significant research contributions.

Let us consider the following simple problem to understand how randomness produces good results:

Problem

Given a set $S = \{a_1, a_2, \cdots, a_n\}$ of n integers, choose an integer in the upper half of S; i.e., it is greater than or equal to more than half of the elements in S.

In order to choose such an integer from S correctly, we need at least $n/2$ comparisons of integers of S. From the following way, $n/2$ comparisons are sufficient to choose such an integer:

1. Let $x = a_1$ be a candidate of such an integer.
2. For each $i(i = 2, 3, \ldots, n/2 + 1)$, compare x and a_i, and if $x < a_i$, then set x to be a_i.

By the deterministic algorithm shown above, $n/2$ is the necessary and sufficient number of comparisons to choose an integer that surely belongs to the upper half of S.

Now, we consider a randomized algorithm for this problem. Randomly choose two integers and compare them. If we answer that the larger one belongs to the upper half of S, then the probability of the correctness is $1 - (1/2)^2 = 3/4$. Although we use only one comparison of integers, the result is not bad. Next, randomly choose one more integer from the rest of S and compare the larger one in the first comparison and the third one. If we answer that the larger one in the second comparison is an integer that belongs to the upper half of S, the probability of the correctness becomes $1 - (1/2)^3 = 7/8$. In this way, we can increase the probability of correctness by increasing the number of comparisons. In general, if we use k comparisons, the probability of the correctness is $1 - (1/2)^{k+1}$. For example, if we use nine comparisons in this way, the probability of an incorrect answer is about $1/1000$. This randomized method is efficient, and we can increase the probability of the correctness as much as we like by simply increasing the computing time.

Randomized algorithms can be divided into two major classes: *Las Vegas* and *Monte Carlo* algorithms. While Las Vegas algorithms never return an incorrect answer, they sometimes may not give any answer at all. Randomized quicksort is an example of a Las Vegas algorithm. Note that randomized quicksort always gives the correct answer, although its running time for the same input may vary considerably from one execution to the next. The name *Las Vegas algorithms* comes from a popular city in Nevada, which is internationally famous for gambling. Las Vegas algorithms were first introduced in 1979 by the Hungarian mathematician Laszlo Babai (1950–) for the graph isomorphism problem.

On the other hand, a Monte Carlo algorithm always gives an answer, but not necessarily a correct one. The probability of success by a Monte Carlo algorithm increases as the algorithm is run repeatedly. Monte Carlo is a part of Monaco, which is surrounded by France and the Mediterranean Sea. It is also well known for its casinos and gambling.

A primality test is an algorithm for determining whether a given number is a prime. Primality testing has applications in many diverse areas, for example, public-key cryptography [8, 9] (see Chapter 30). A naïve primality test is as follows: Given an input number, check whether any integer from 2 up to the square root of n can be a factor of n. If n is a composite, then it can be factored into two integers. This naïve method takes an exponential time in log n for input n. Most popular and practically efficient primality tests are probabilistic ones. The Miller–Rabin primality test (1976; Gary Lee Miller) and Solovay–Strassen primality test (1977; Robert M. Solovay, 1938–; Volker Strassen, 1936–) are sophisticated Monte Carlo algorithms that usually produce good results. The running time of these algorithms is $O((\log n)^3)$.

We can consider some complexity classes based on randomized algorithms. The following are examples of well-known classes [15]:

1. The class **ZPP** (for zero-error probabilistic polynomial) is the class of languages that have Las Vegas algorithms running in expected polynomial time.

2. The class **RP** (for randomized polynomial) is the class of languages L that have a randomized algorithm A running in polynomial time such that for any input x:

    a. If x is in L, then the probability that x is accepted by A is at least $1/2$.

    b. If x in not in L, then the probability that x is accepted is 0.

3. The class **BPP** (for bounded-error probabilistic polynomial) is the class of languages L that have a randomized algorithm A in polynomial time such that for any input x:

    a. If x is in L, then the probability that x is accepted by A is at least $3/4$.

    b. If x is not in L, then the probability that x is accepted by A is at most $1/4$.

Randomized quicksort is a **ZPP** algorithm. There are several interesting open problems regarding the relationships among randomized complexity classes, for example, the question of whether **BPP** $\subseteq$ **NP** is open.

REFERENCES

1. A. V. Aho, J. E. Hopcroft, and J. D. Ullman, *The Design and Analysis of Computer Algorithms*, Addison-Wesley, Reading, MA, 1974.
2. D. E. Knuth, *The Art of Computer Programming: Fundamental Algorithms* (vol. 1, 2nd ed.), Addison-Wesley, Reading, MA, 1973 (the first edition was published in 1968).
3. D. E. Knuth, *The Art of Computer Programming: Seminumerical Algorithms* (vol. 2, 2nd ed.), Addison-Wesley, Reading, MA, 1981.

4. D. E. Knuth, *The Art of Computer Programming: Sorting and Searching* (vol. 3), Addison-Wesley, Reading, MA, 1973.
5. N. Wirth, *Algorithms + Data Structures = Programs*, Prentice-Hall, Englewood Cliffs, NJ, 1976.
6. G. Brassard and P. Bratley, *Algorithmics: Theory and Practice*, Prentice-Hall, Englewood Cliffs, NJ, 1988.
7. U. Manber, *Introduction to Algorithmics: A Creative Approach*, Addison-Wesley, Reading, MA, 1989.
8. T. H. Cormen, C. E. Leiserson, and R. L. Rivest, *Introduction to Algorithms*, MIT Press, Cambridge, MA, 1990.
9. J. Kleinberg and E. Tardos, *Algorithm Design*, Addison-Wesley, Boston, 2006.
10. Wikipedia, Tony Hoare, http://en.wikipedia.org/wiki/Tony_Hoare.
11. Wikipedia, Quick Sort, http://en.wikipedia.org/wiki/Quick_sort.
12. Wikipedia, Linked List, http://en.wikipedia.org/wiki/Linked_list.
13. M. L. Fredman and R. E. Tarjan, Fibonacci Heaps and Their Uses in Improved Network Optimization Algorithms, *Journal of the ACM*, 34, 596–615, 1987.
14. E. W. Dijkstra, A Note on Two Problems in Connection with Graphs, *Numerische Mathematik*, 1, 269–271, 1959.
15. R. Motwani and P. Raghvan, *Randomized Algorithms*, Cambridge University Press, New York, 1995.

CHAPTER 28

Parallel and Distributed Computing

28.1 DAWN OF PARALLELISM

Parallel computing is a form of computation in which more than one calculation can be concurrently carried out. A parallel computer is a computer system with multiple processing elements, working in parallel, to solve a problem. Before the middle of the 1950s, all commercial computers were traditional serial computers.

IBM 704 was introduced in 1954. It was the first mass-produced computer with floating-point arithmetic hardware and could execute up to 4000 instructions per second. IBM 704 was a very successful commercial computer. However, after the middle of the 1950s some research projects needed much faster computers. For example, the University of California Radiation Laboratory (UCRL) in Livermore, California, and Los Alamos Scientific Laboratory (LASL) wanted high-performance computers for their projects. In April 1955, IBM submitted a proposal to UCRL, but UCRL rejected it, instead getting in contact with Remington Rand (UNIVAC). Then IBM submitted a proposal of STRETCH (also known as IBM 7030) to LASL in 1956, and was awarded the contract with LASL for the high-performance computer system.

STRETCH was an amazing computer system in the 1950s that contained many high-performance features, such as local concurrency, nonlocal concurrency, multiprogramming, a look-ahead approach to start memory fetches early, and pipeline utilization. John Cocke (1925–2002) contributed to developing these ideas. From these features, we can say that STRETCH was an aggressive computer system with single-processor parallelism. We may therefore consider that the start of the STRETCH project is the dawn of parallelism. The STRETCH design had its roots in 1954 from initial studies on advanced concepts for high-performance computing by Stephen W. Dunwell (1913–1994) and Werner Buchholz (1922–). The STRETCH project started formally in 1955 after UNIVAC won the contract to build the Livermore Automatic Research Computer (LARC). After losing the competition on LARC, IBM proposed a high-performance computer system that was 100 times faster than that of IBM 704 to the Los Alamos Scientific Laboratory in 1955. John Cocke won the Turing Award for his large contribution to computer architecture and compiler optimization in 1987.

In 1961, actual benchmarks indicated that the performance of the IBM 7030 was only about 30 times faster than that of the IBM 704. While the IBM 7030 was not considered successful, it spawned technologies incorporated in future computer systems. The STRETCH was conceived as a *supercomputer* since its high-performance and new concepts of advanced technology were far beyond the level of existing computer systems in the 1950s. Many advanced technologies developed with the STRETCH project were incorporated in later supercomputer designs, such as IBM System/360 models, IBM System/370 models, and the IBM 3090 series. As the editor, Werner Buchholz published a book about the STRETCH project in 1962 [7]. He is the person who coined the term *byte* in 1956, a unit of digital information (1 Byte = 8 bits). The first STRETCH was delivered to Los Alamos Scientific Laboratory (LASL) in 1961, and used until 1971. The second STRETCH was delivered to the U.S. National Security Agency as part of the HARVEST system in 1962. Altogether, 8 STRETCH systems (six in the United States, one in the UK, and one in France) were sold from 1961 to 1963.

Frances E. Allen (1932–) joined IBM in 1957 and ended up staying there for 45 years. Her work has had strong impacts on compiler research and practice. She introduced many of the abstractions, algorithms, and implementations that laid the groundwork for automatic program optimization technology. Allen developed and implemented her methods as part of the compiler for the IBM STRETCH-HARVEST system. She became the first woman to win the Turing Award in 2006.

28.2 PARALLEL COMPUTERS

Physical limitations on processing speeds forced high-performance computations to be targeted at the exploitation of parallelism. Parallel computer architecture has grown in the form of multiple microprocessors.

Daniel L. Slotnick (1931–1985) studied mathematics at Columbia University and New York University (now called the Courant Institute at New York University). In 1957, he joined IBM, where he wrote a joint paper with John Cocke on the use of parallelism in numerical calculations. Slotnick was then employed by the Westinghouse Electric Corporation in Baltimore, Maryland, where he was given the opportunity to pursue his ideas on parallel computers. He designed the Solomon computer and built the first parallel processor prototype using first a 3 by 3 and then a 10 by 10 processor array. In the 1960s, 258 processor elements were used in the Solomon computer. These processor elements could run a single instruction at a time in parallel. The concept of applying a single instruction to a large number of data elements is now commonly referred to as single instruction, multiple data (SIMD). In 1964, a prototype of Solomon was built under a contract from the U.S. Air Force, but the contract ended and Westinghouse gave up developing the Solomon system any further.

In 1965, Slotnick moved to the University of Illinois at Urbana-Champaign and started the ILLIAC IV project with Burroughs under the sponsorship of the government Advanced Research Project Agency (ARPA). Among its technological innovations, ILLIAC IV was the first large computer system that employed semiconductor primary memory. It was a SIMD computer for array processing. The ILLIAC IV design featured high parallelism with up to 256 processors that were used to allow it to work on large data sets. That technique would later be known as *vector processing*. After many problems, in 1970, ILLIAC IV was transferred from the campus of the University of Illinois at Urbana-Champaign to Moffett Field, California, NASA Ames Research Center. The first run of ILLIAC IV was in 1973, but it was not fully operational until 1975. Its performance was about 200 MFLOPS (200 million floating-point operations per second) and its clock frequency was 13 MHz. The ILLIAC IV was credited as the fastest computer in the world until 1981. The operation of ILLIAC IV was eventually ended in 1982. The ILLIAC IV chassis is now displayed at the Computer History Museum in Mountain View, California.

Cray-1 was the first commercially successful supercomputer and was manufactured by Cray Research, Inc., founded in 1972 by Seymour Roger

Cray (1925–1996). A *supercomputer* is a computer that is at the most advanced frontline of current processing capacity. The architecture of Cray-1 was designed mainly by Seymour Cray. The first Cray-1 was lent to Los Alamos National Laboratory in 1976 for a 6-month trial. The first full system of Cray-1 was sold to the National Center for Atmospheric Research (NCAR) in 1977. Cray-1 adopted integrated circuits (ICs). About 200,000 gates were used in Cray-1, and these ICs were supplied by Fairchild Semiconductor and Motorola. Over 80 Cray-1 systems were sold, and the company was very successful in the supercomputer market. When Cray-1 was released, it beat almost every computer in computing speed. Only ILLIAC IV was nearly at the same level of performance, but its operational cost was much behind that of Cray-1. The peak speed of the first Cray-1 was 250 MFLOPS.

Cray-1 was succeeded in 1982 by the 800 MFLOPS Cray X-MP, the first Cray multiprocessing computer. In 1985, the very advanced Cray-2 appeared. Its peak performance was at first 1.9 GFLOPS (1.9×10^9 floating-point operations per second), and improved to 3.9 GFLOPS. Cray-1s are now on display at a number of museums (e.g., Computer History Museum in Mountain View, California, Science Museum in London, and Deutsches Museum in Munich).

In the early and mid-1980s, a standard supercomputer was a computer system with a modest number of vector processors, typically in the range of 4 to 16, working in parallel. In the later 1980s and 1990s, a supercomputer became a massive parallel processing system with a thousand or more processing units. Traditionally, U.S. computer companies such as Cray, IBM, and Intel had dominated in the supercomputer market. In the 1990s, Japanese computer companies NEC, Fujitsu, and Hitachi came up to the top group in the supercomputer market, but in the late 1990s, Hitachi and Fujitsu moved down from the top group, and in 2002, NEC lost its top position to IBM Blue Gene/L. Most modern supercomputers are now highly tuned computer clusters using commodity processors combined with custom requests. In 2010, Tianhe-1A at National Supercomputing Center, Tianjin, China, became the fastest supercomputer in the world. The speed of Tianhe-1A is 2.5 PFLOPS (2.5×10^{15} floating-point operations per second), whereas the speed of the second fastest supercomputer, Cray Jaguar, is 1.76 PFLOPS. In 2011, the *K* computer, manufactured by Fujitsu, became the world's fastest supercomputer with a computation speed of 8 PFLOPS, but in 2012 the IBM Sequoia became the world's fastest supercomputer. The K computer, currently installed in Kobe, Japan, is the fourth fastest supercomputer in the world as of 2013 [8].

Supercomputers are used for scientific research in such fields as weather forecasting, climate research, quantum physics, molecular modeling, biological macromolecules, and physical simulation, but also for intelligence and military uses.

28.3 PARALLEL ALGORITHMS

A parallel algorithm is a procedure that executes pieces of work at the same time on many different processing devices. Some algorithms are easily divided into pieces of work to be allocated to different processors, but some algorithms are not easy to do so. It is convenient for designing an efficient parallel algorithm if we have a suitable model of the parallel computer.

A widely accepted model for designing and analyzing sequential algorithms consists of a central processing unit with a random access memory (RAM) attached to it. This model is also called the von Neumann model. The RAM model has been successful in estimating the performance of sequential algorithms. We can consider that it is an efficient and useful bridge between software and computer hardware. It is not easy to give a commonly accepted model for parallel computation due to the presence of a set of interconnected processors and their concurrency. The performance of a parallel algorithm usually depends on various factors, such as processor allocation, job scheduling, communication, and synchronization among processors working concurrently.

One of the commonly used models for parallel computing has been the shared memory model. It consists of a number of processors, each of which has its own local memory and can execute its own local program. All of the processors can communicate by exchanging data through a shared memory unit. There are two basic types of the shared memory model. If all the processors operate synchronously under the control of a common clock, the model is called the parallel random access machine (PRAM) model. The other type is called the asynchronous model. In the asynchronous model, each processor operates under its own clock. A PRAM is considered a parallel computer that operates multiple instructions on multiple data (MIMD type). That is, each processor may execute multiple instructions on data different from those executed by any other processor during any given time unit. A general view of a shared memory model with n processors is shown in Figure 28.1.

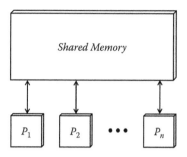

FIGURE 28.1   The shared memory model.

The network model has also been widely used for parallel computing. In particular, a network model may be suitable in the case where the communication costs among processors are considered to be large. A network can be viewed as a graph $G = (V, E)$, where each node in V represents a processor and each edge (i, j) in E represents a communication link between processors i and j. Each processor is assumed to have its own local memory and control unit, and no shared memory is available. Data can be exchanged between two processors through the communication link. The network model incorporates the topology of the interconnection between the processors into the model itself. The linear processor array, the tree-connected array, the two-dimensional mesh (it is also called the mesh-connected processor array), and the hypercube are examples of widely used topologies. An example of the mesh-connected processor array is shown in Figure 28.2.

There is a very large body of literature on the subject of PRAM algorithms and network model algorithms. A lot of parallel algorithms have appeared since the middle 1970s until today. Some of them are very smart, and some of them are very sophisticated. Here, we describe two fundamental examples of parallel algorithms.

> **Parallel merge sorting.** The parallel merge sort is based on a merging procedure that is used to sort successively larger and larger non-overlapping subsequences until the whole sequence is sorted. The sequence of operations by the parallel merge sort algorithm can be represented on a binary tree as follows. Let T be a balanced binary tree with n leaves. The elements of the sequence are distributed among leaves. The nodes at height 1 represent the lists that we obtained by merging the pairs of consecutive elements contained in the children nodes. In general, each internal node represents the sequence that

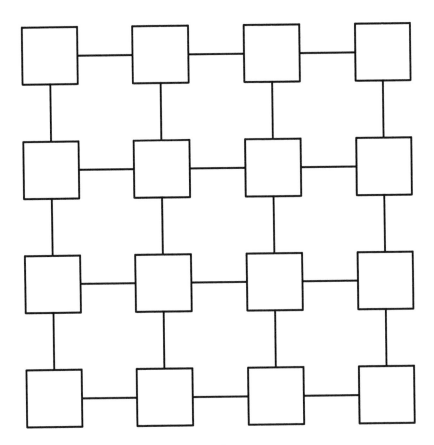

FIGURE 28.2   The 4 × 4 mesh-connected processor array.

we obtained by merging the subsequences generated at the children nodes. Hence, each internal node represents the sorted list of the elements sorted in the subtree rooted at the node. If we use the optimal O(log log n) time merging procedure proposed by Leslie G. Valiant (1975), we obtain an O(log (log log n)) time parallel sorting algorithm. The total number of operations of the algorithm is O(n log n). The pipelined merge sorting algorithm, called Cole's merge sort, was developed in 1988 by Richard Cole (1956 –).

List ranking. The list ranking problem is often encountered in the design of parallel algorithms as the fundamental technique. It is defined as follows: given a linked list L with n nodes, we would like to compute an array R such that $R(i)$ is the distance of node i from the end of L. This problem was first proposed in 1979 by J. C. Wylie (1952 –). The following is a simple parallel list ranking algorithm:

```
begin
for i := 1 to n do in parallel
   if s(i) ≠ 0 then R(i) := 1
   else R(i) := 0;
for i := 1 to n do in parallel
   begin
      q(i) := s(i)
      while q(i) ≠0 and q(q(i)) ≠0 do begin
      R(i) := R(i) + R(q(i));
      q(i) := q(q(i)) end
   end
end;
```

The simple list ranking algorithm shown above generates $R(i)$ of each node i in $O(\log n)$ time using $O(n \log n)$ operations. The algorithm can be implemented to run on the PRAM model. The optimal list ranking algorithm that takes $O(\log n)$ time and $O(n)$ operations was first discovered by Richard Cole and Uni Viskin (1952–).

The computational complexities of parallel algorithms have also been much studied. In the case of sequential algorithms, polynomial time solvable problems are considered to be feasible. For parallel algorithms, we use the different definition for the quickly solved standard. The complexity class, Nick's class (NC), of problems quickly solved on a parallel computer was named by Stephen Cook after Nicholas Pippenger (1946–), who had done extensive research on circuit complexity. If a problem is solved by a Boolean circuit with polylogarithmic depth ($O(\log^c n)$) and polynomial size ($O(n^k)$) for some constants c and k, we say that the problem belongs to NC^c. It is obvious that $NC^1 \subseteq NC^2 \subseteq \ldots \subseteq NC^c \subseteq \ldots$. Nick's class NC is defined to be $NC^1 \cup NC^2 \cup \ldots \cup NC^c \cup \ldots$. In other words, if a problem belongs to NC, it can be solved in time $O(\log^c n)$ using $O(n^k)$ parallel processors for some constants c and k. One of the major open problems is whether or not every class containment in the NC hierarchy is proper. That is, it is open to whether the following proper inclusion relation holds true:

$$NC^1 \subset NC^2 \subset \ldots \subset NC^c \subset \ldots \subset NC.$$

In 1994, Christos H. Papadimitriou (1950–) showed that $NC^1 \subseteq L \subseteq NL \subseteq NC^2$, where L (also known as LSPACE) is the complexity class containing decision problems that can be solved by a deterministic Turing machine using a logarithmic amount of memory space. NL is the complexity class containing decision problems that can be solved by a nondeterministic Turing machine using a logarithmic amount of memory space.

More material about parallel algorithms can be found, for example, in [1, 2].

28.4 DISTRIBUTED COMPUTING

A distributed system originally referred to computer networks where individual computers were physically distributed within some geographical area. The term *distributed system* is now used in a wider sense. Even if some autonomous processors run on the same physical computer and they interact with each other by message passing, we may consider such a system a distributed system. The study of distributed algorithms has its roots in designing operating systems for distributed systems in the 1960s.

The asynchronous network model and the asynchronous shared memory model are widely used in the area of distributed computing. A distributed system is a collection of individual computing devices called processes or processors together with communication channels. Processes can communicate with other processes through communication channels in a network or through the shared memory as a communication model. The shared memory is an abstraction of asynchronous interprocess communication. Each process in a distributed system is generally performed by its own program, but it is occasionally requested to collaborate with other processes. In the following, we mainly describe distributed computing on the shared memory model.

Interaction between a process and its corresponding user is by input actions from the user to the process and by output actions from the process to the user. We may consider each process in the distributed system to be a state machine. All communication among the processes is via the shared memory (also called variables). This model was introduced by Nancy A. Lynch (1948–) and Mark R. Tuttle (1962–) in 1987, and it is known as I/O automaton [2]. Two types of the shared memory model have been widely used. One is the multiwriter/reader shared memory, and the other is the single-writer/multireader shared memory. In the multiwriter/reader shared memory model, the same shared variable may be read or written by different processes. On the other hand, in the single-writer/multireader shared memory model, each shared variable can be written by only one process, but may be read by any process.

The behavior of operational executions in a distributed system should be required to be consistent for all processes and interprocess communication. We therefore need a unified theory of shared memory consistency. In 1986, Leslie Lamport (1941–) defined three categories, *safe*, *regular*, and *atom*, for shared variables according to possible assumptions about what can happen in the concurrent case of read operations and write operations [9]. A shared variable is safe if every read operation that does not overlap with write operations returns the last value written to the shared variable. On the other hand, every read operation that overlaps with one or more write operations may return any value from the domain of the shared variable. A shared variable is said to be atomic if it is regular with the additional property that read operations and write operations behave as if they occur in some total order. In a distributed system, many distributed algorithms have been designed under the assumption that all shared variables are atomic. That is, in the design of most distributed algorithms, we assume that there is a possible linearization of the temporal order of read operations and write operations such that the linearization is consistent with the actual behavior of the system, although these operations may be physically overlapped. Even if different processes try to write on the same shared variable at nearly the same time, one process's writing precedes the other process's writing. This means that the contents of the shared variable by the earlier one are changed to the value written by the later one even if these two events occur very closely.

Mutual exclusion is one of the most fundamental problems for distributed computing. Historically, it was first seriously studied in 1965 by Edsger W. Dijkstra (1930–2002) as an important problem for a distributed operating system [10]. It is the problem of how to allocate a single individual, nonshareable resource among users. A user with access to the resource is modeled as being in a critical region (i.e., admitted state to use the resource). When a user is not involved in any way with the resource, it is said to be in the remainder region. In order to gain admittance to its critical region, a user executes a trying protocol. The duration from the state of executing the trying protocol to the entrance of the critical region is called the trying region. After the end of the use of the resource by a user, it executes an exit protocol. The duration of executing the exit protocol is called the exit region. Each user follows a cycle, moving from its remainder region to its trying region, then to its critical region, then to the exit region, and finally back to its remainder region. This cycle can be repeated.

The mutual exclusion problem is to design a fair and efficient algorithm to decide the temporal order among users wishing to use a shared resource. The distributed system to solve the mutual exclusion problem should satisfy the following conditions:

1. There is no reachable system state in which more than one user is in the critical region.

2. If at least one user is in the trying region and no user is in the critical region, then at some later time point some user enters the critical region.

3. If a user is in the exit region, then at some later time point some user enters the remainder region.

If a mutual exclusion algorithm satisfies the following two additional conditions, it is said to be lockout-free:

4. If all users always return the resource, then any user that reaches the trying region eventually enters the critical region.

5. Any user that reaches the exit region eventually enters the remainder region.

An early algorithm for the mutual exclusion by Dijkstra guarantees mutual exclusion, but it does not guarantee lockout freedom. That is, Dijkstra's algorithm may allow one user to be repeatedly granted access to its critical region, while other users trying to gain access never succeed in doing so. Subsequently, a number of improved mutual exclusion algorithms have been proposed.

Other typical problems in a distributed system are the *leader election problem*, *consensus problem*, *resource allocation problem*, *synchronizer construction problem*, *concurrent snapshot problem*, and *bounded time-stamp problem*. These problems have been extensively studied in distributed environments since the 1970s. More material about distributed computing and algorithms can be found, for example, in [3–6].

REFERENCES

1. J. Jaja, *An Introduction to Parallel Algorithms*, Addison-Wesley, Reading, MA, 1992.
2. F. T. Leighton, *Introduction to Parallel Algorithms and Architectures*, Morgan Kaufmann, San Mateo, CA, 1992.

3. G. Tel, *Topics in Distributed Algorithms*, Cambridge University Press, New York, 1991.

4. G. Tel, *Introduction to Distributed Algorithms*, Cambridge University Press, New York, 1994.

5. N. A. Lynch, *Distributed Algorithms*, Morgan Kaufmann, San Francisco, 1996.

6. H. Attiya and J. Welch, *Distributed Computing* (2nd ed.), John Wiley & Sons, Hoboken, NJ, 2004.

7. W. Buchholz, *Planning a Computer System: Project STRETCH*, McGraw Hill, New York, 1962.

8. Wikipedia, K Computer, http://en.wikipedia.org/wiki/K_computer.

9. L. Lamport, The Mutual Exclusion Problem, Part II, *Journal of the ACM*, 33, 327–348, 1986.

10. E. W. Dijkstra, Solution of a Problem in Concurrent Programming Control, *Communications of the ACM*, 8, 569, 1965.

CHAPTER 29

Computer Networks

Sputnik 1 was the first Earth-orbiting artificial satellite launched by the Soviet Union on October 4, 1957. Its success was an astonishing shock to the United States. This incident is considered the start of the Space Race during the Cold War. *Explorer 1* was the first Earth-orbiting artificial satellite launched by the United States on January 31, 1958. Both the United States and the Soviet Union thought that the development of space technology could provide a huge diplomatic and military advantage [3–5].

In February 1958, the Advanced Research Projects Agency (ARPA) was established in the U.S. Department of Defense. Its foundation was motivated by American competition with the Soviet Union's launch of *Sputnik 1*. The United States was trying to develop a defense system against ballistic missile attacks. ARPA promoted research projects for improving U.S. military technology. In 1972, ARPA was renamed the Defense Advanced Research Projects Agency (DARPA), and then in 1993 reverted to ARPA. It was finally renamed DARPA in March 1996 [6].

29.1 PACKET SWITCHING NETWORKS

On May 28, 1961, the radio relay stations at Cedar Mountain, Utah, and Wendover, Nevada, were blown up by radicals and virtually destroyed. More than 2200 telephone and telegraph circuits and two television channels were interrupted. The U.S. government was urged by this incident to develop secure and reliable communication systems that would be invulnerable to communication facility damage. The U.S. government transferred its Command and Control Project from Defense Research Engineering to ARPA. In the same year, ARPA commissioned the RAND

Corporation to conduct a research project for the construction of invulnerable communication systems that could tolerate nuclear missile attacks. The RAND Corporation (Research and Development Co.) is a nonprofit global policy think tank that was formed in 1946 to conduct research and analysis for the U.S. armed forces [9].

Paul Baran (1926–2011) was a Polish-born American engineer who joined the RAND Corporation in 1959 to work on a project to design a robust communication system that could maintain communication between endpoints even in the event of serious damage from explosions or attacks. The concepts of packet switching were first explored by Baran in the early 1960s. Independently, Donald Davies (1924–2000) at the National Physical Laboratory in the UK and Leonard Kleinrock (1934–) at MIT also developed similar ideas [8].

Packet switching is a digital networking communication method that groups all transmitted data regardless of content, type, or structure into suitably sized blocks called packets. The principal goals of packet switching are to optimize the utilization of available link capacity, minimize response times, and increase the robustness of communication against the damage of nodes or lines in a communication network [2]. Baran may have obtained his ideas for packet switching techniques from the observation of the similarity between biological neural networks and communication systems for telephones and telegraphs. Baran and his team at the RAND Corporation developed simulation tests of the connectivity among nodes in a communication network. A network with at least three connected links at each node showed a significant increase in resilience even when many nodes were eliminated. Details of the designs of robust communication networks using packet switching techniques were reported in "On Distributed Communication," which was published by the RAND Corporation and submitted to ARPA in 1964 [10, 11].

Data communications based on the idea of circuit switching, such as in traditional telephone circuits, were possible only between the two connected parties. With packet switching, however, data packets from one party could be transmitted to many different destinations, and each data packet could be routed independently [2].

29.2 ARPANET AND CSNET

Joseph Carl Robnett Licklider (1915–1990) studied physics and mathematics (B.A. in 1937) and psychology (M.A. in 1938) at Washington University in St. Louis. He received his Ph.D. in psychoacoustics from Rochester

University in 1942. Around 1950 he became interested in information technology, and in August 1962 he conceived his earliest ideas for a computer network allowing communications among computer users. Licklider wrote a series of memos about his concept of computer networks, where everyone in the world could be connected and access programs and data at any site from anywhere. He called his computer network concept the *Intergalactic Computer Network* (*Galactic Network* for short) [14].

In 1962, Licklider became the head of the computer research group at ARPA, called the Information Processing Technology Office (IPTO), where he developed his ideas on how to establish a time-sharing network of computers. He discussed the Galactic Network with young computer scientists at MIT. In 1963, Licklider started to discuss his vision with Laurence G. Roberts (1937–), Ivan Sutherland (1938–), and Robert Taylor (1932–). Licklider contracted with MIT, UCLA, and BBN Technologies to start working on computer networks. Eventually, his vision of computer networks led to the establishment of the ARPANET, the world's first operational packet switching network. Today, Licklider is remembered as one of the pioneers of the Internet [12, 14].

Ivan Sutherland used the TX-2 at MIT to write graphical programs for computer-aided design. He received the Turing Award in 1988 for the invention of *Sketchpad*, an early predecessor to a sort of graphical user interface. Sutherland became the head of IPTO at ARPA when Licklider returned to MIT in 1964. Robert Taylor (1932–) was appointed the head of IPTO at ARPA when Sutherland moved to Harvard University in 1965.

The ARPANET project actually started in December 1966. Robert Taylor had three computer terminals, each connected to different computers: the first for the System Development Corporation (SDC) Q-32 in Santa Monica, the second for Project Genie at the University of California, Berkeley, and the third for Multics at MIT. Taylor thought that rather than having three terminals, there ought to be just one terminal capable of connecting anywhere one would want to communicate. That idea was the fundamental concept of the ARPANET [12].

In 1968, Robert Taylor proposed a plan for a computer network composed of small computers, called interface message processors (IMPs). The IMP at each node would work as a gateway (router) and would perform the store-and-forward packet switching function. The host computers were connected to the IMPs via a communication interface. The first-generation IMPs were initially built by BBN Technologies using Honeywell DDP-516 computers. Each IMP could support up to four local host computers, and could communicate with up to six remote IMPs. The initial ARPANET consisted of the IMPs at the University of California, Los Angeles (UCLA), Stanford Research Institute, University of California, Santa Barbara (UCSB), and University of Utah [2, 12].

On October 29, 1969, the first message on the ARPANET was transmitted from the SDS Sigma 7 host computer at UCLA to the SDS 940 host computer at Stanford Research Institute. By December 5, 1969, the entire four-node network (the IMPs at UCLA, Stanford Research Institute, UCSB, and the University of Utah) was established. Thereafter, each year the ARPANET grew significantly, and by 1981, the number of host computers connected to the ARPANET exceeded 200 sites [12].

Robert Elliot Kahn (1938–) and Vinton Cerf (1943–) are American computer scientists who invented the *Transmission Control Protocol* (TCP) and the *Internet Protocol* (IP), currently the fundamental communication protocols for the Internet. The set of these protocols is referred to as TCP/IP. Kahn, who began to work at IPTO in ARPA in 1972, demonstrated the ARPANET by connecting 20 different computers at the International Computer Communication Conference. Thereafter he worked to develop the TCP/IP protocols. In 1973, Vinton Cerf joined Kahn's team on the TCP project, and they completed an early version of TCP. Later it was separated into two layers, where the most fundamental functions of TCP were moved to the Internet Protocol (IP). Thus, TCP works as an intermediary level between an application program and the Internet Protocol. Kahn and Cerf received the Turing Award in 2004 for their pioneering work on the Internet [12, 15, 16].

The CSNET was established in Madison, Wisconsin, in 1979 by Lawrence Landweber (1941–). He invited a group of colleagues from other universities as well as representatives from DARPA and the National Science Foundation (NSF) to discuss the possibility of constructing a computer network connecting computer science departments. MIT, Carnegie Mellon University, Stanford University, and some other major universities already used the ARPANET. However, many participants at the meeting in Madison were not affiliated with these major universities. They thought that computer network access would be very important for all computer science departments and believed that computer network communication could significantly improve the research environment for scientists.

Landweber organized an electronic mail facility for theoretical computer scientists called THEORYNET. It provided members with a mailbox on a central computer at the University of Wisconsin. The members accessed it from terminals over dial-up phone lines or through the Telnet public packet-switched network. THEORYNET successfully attracted many users, and as a result, Landweber wanted to extend THEORYNET to include file transfer, remote login, and faster message delivery. He extended the original CSNET proposal to a consortium of universities. The CSNET proposal gained the support of DARPA and NSF in 1980. By 1981, three sites (the University of Delaware, Princeton University, and Purdue University) had joined CSNET. By 1982, 24 sites had joined, which expanded to 84 sites by 1984, including one in Israel. Soon thereafter, connections were further expanded to computer science departments in Australia, Canada, France, Germany, Japan, and Korea. During this period, a gateway node was installed at the University of Wisconsin to provide access to the ARPANET. The ARPANET and CSNET were the forerunners of the Internet [13].

29.3 WORLD WIDE WEB

The *World Wide Web* (WWW; commonly known as the Web) is a system of interlinked hypertext documents that can be accessed via the Internet. It rapidly became popular among Internet users in the 1990s.

The conceptual groundwork for the World Wide Web started with an American engineer and science administrator, and one of the creators of the National Science Foundation, Vannevar Bush (1890–1974), who introduced a conceptual machine called the *Memex* (derived from "memory extension") during the 1940s. Bush imagined a microfilm-based device in which all his books, documents, and records could be stored. It would be mechanized for high-speed searching and flexibility. The Memex would serve as a supplement to the user's memory, where private files and data were stored in such a way that any item could lead to another related item quickly, much like the way hyperlinks work on the Internet. He wrote about the Memex in 1945 in a seminal article entitled "As We May Think" published in the *Atlantic Monthly*. In this article, he described his idea as an adjustable microfilm viewer. The Memex would work as a memory bank to organize and retrieve data through the use of "associative trails." It was somewhat analogous to the structure of some present-day databases and that of the present World Wide Web [18].

Theodor Holm Nelson (1937–), an American sociologist, philosopher, and pioneer of information technology, coined the terms *hypertext* and *hypermedia* in 1963 to describe the new paradigms for building tools that would transform our way of reading and writing. He published his ideas in the article "A File Structure for the Complex, the Changing, and the

Intermediate" for the 20th National Conference of ACM in 1965. Nelson thought that the concepts of Bush's Memex could be better applied to computer networks than to photoelectrical or mechanical devices. The main thrust of his work has been to make information easily accessible to ordinary people. Nelson founded Project *Xanadu* in the 1960s, intending to create a computer network with a simple user interface. The project name *Xanadu* came from the poem "Kubla Khan" by Samuel Taylor Coleridge (1772–1834). Xanadu was the summer capital of Kublai Khan's Yuan Empire, which Coleridge described as a dreamland. The poem suggested to Nelson an image of a vast storehouse of memory [19].

In 1974, Nelson published his book *Computer Lib/Dream Machine*, where he defined *hypertext* as nonsequential writing. Ordinary writing is sequential for two reasons. First, it grows out of speech and speech making, and second, because books can only be read conveniently in sequence. A footnote is a break from a sequence, but it cannot really be extended. Writers do better if they do not have to write sequentially, and readers do better if they do not have to read in an imposed sequence, but may establish impressions, jump around, and try different pathways until they find the ones they want to follow and study most closely. Nelson devoted much of his time to working on and advocating for Project Xanadu. He intended to establish an ideal new publishing system for the digital age, although Project Xanadu was not quite successful from a practical viewpoint. His visionary design was later realized by the invention of the World Wide Web by Tim Berners-Lee (1955–) in 1989.

Tim Berners-Lee, an English engineer and computer scientist, studied at Queen's College, University of Oxford, from 1973 to 1976. While working for CERN (Conseil Europeen pour la Recherche Nucleaire, or the European Organization for Nuclear Research) from June to December 1980, he proposed a project based on the concept of hypertext to facilitate sharing and updating information among researchers. During that period, he built a prototype system named ENQUIRE, which allowed links to be made between arbitrary nodes in a computer network [7, 20].

In 1989, Berners-Lee had an opportunity to join hypertext with the Internet. He wrote his initial proposal in 1989 for what would eventually become the World Wide Web. In 1990, Berners-Lee and a Belgian computer scientist, Robert Cailliau (1947–), produced a revised version of the proposal. They used similar ideas to those in the ENQUIRE system. Then they designed and built the first Web browser with the function of an editor, and the first Web server. Berners-Lee's breakthrough was to combine hypertext with the Internet. As described in his book *Weaving the Web: The Original Design and Ultimate Destiny of the World Wide Web by Its Inventor* [21], he suggested that a marriage between the two technologies was possible to members of both technical communities. In the process of working on his project, Berners-Lee developed three essential technologies [17, 20, 21]:

1. A system of globally unique identifiers for resources on the Web and elsewhere, the Universal Document Identifier (UDI), later known as Uniform Resource Locator (URL) and Uniform Resource Identifier (URI)

2. The publishing language Hypertext Markup Language (HTML)

3. The Hypertext Transfer Protocol (HTTP).

In 1993, CERN announced that the World Wide Web would be free to anyone, requiring no fees due. Since the World Wide Web was non-proprietary, it was possible to develop servers and clients independently and add extensions without licensing restrictions. In 1994, the World Wide Web Consortium (W3C) was founded by Berners-Lee in the Laboratory for Computer Science at MIT (LCS/MIT) with support from DARPA. A year later, a second website was founded at IRIA (the French national computer science laboratory) with support from the European Commission DG (Directorates—General Information Society of the European Commission). By the end of 1994, the total number of websites was still quite small compared to the present. Since the mid-1990s, the number of websites has increased rapidly. Connected by the existing Internet, websites were created around the world, and international standards for domain names and HTML were added. The World Wide Web enabled the spread of information over the Internet through an easy and flexible format, playing an important role in popularizing the Internet for use by both scientists and nonscientists alike [17].

29.4 CLOUD AND GRID COMPUTING

With the development of computer networks, the client-server model of computing was born. Cloud computing is a technique for constructing an infrastructure for shared services. It is primarily used to sell application services running client-server software at remote locations [27, 28]. The following quotation is from "CG Technologies, Cloud Computing" [28]:

Any computer or web-friendly device connected to the Internet may access the same pool of computing power, applications, and files in a cloud-computing environment. Users may remotely store and access personal files such as music, pictures, videos, and bookmarks; play games; or do word processing on a remote server. Data is centrally stored, so the user does not need to carry a storage medium such as a DVD or USB flash drive. Desktop applications that connect to internet-host email providers may be considered cloud applications, including web-based email services and many others.

In the mid-2000s, Amazon became famous for its success in online retailing. In 2002, Amazon Web Services provided a suite of cloud-based services as an online retailer. In 2006, it launched its Elastic Compute Cloud (EC2) as a commercial Web service allowing small companies and individuals to use the cloud computing system to run their own computer applications. In 2007, Google, IBM, and a number of universities embarked on large-scale cloud computing research projects.

The term *grid computing* originated in the early 1990s as a metaphor for making access to distributed computing systems as easy as access to outlets of electric power. In 1998, Carl Kesselman and Ian Foster defined *grid computing* in their book *The Grid: Blueprint for a New Computing Infrastructure* (see also [22]). They described grid computing as follows: "A computational grid is a hardware and software infrastructure that provides dependable, consistent, pervasive, and inexpensive access to high-end computational capabilities." Already in 1969, Leonard Kleinrock presciently suggested a similar concept: "We will probably see the spread of computer utilities, which, like present electric and telephone utilities, will service individual homes and offices across the country" [25].

Grid computing is a form of distributed parallel computing whereby a super and virtual computer is composed of a cluster of networked or coupled computers acting together to perform very large computational tasks. Its main goal is the development of high-performance distributed computing software allowing users to access distributed computing environments such as meta-computing or cluster computing and to produce smart applications to use resources that are geographically separated across large networks. The increasing network bandwidth, more powerful and faster computer processors, and proliferation of Internet technologies have brought a new and better way of computing via the grid concept. This is a research infrastructure that supports computation-intensive and data-intensive collaborative activities through dynamically collected and integrated shared research resources connected by a high-speed network. Many academic institutes, research organizations, and commercial enterprises have been trying to take advantage of this type of computing paradigm, and are constantly seeking new technologies and applications that have not been able to provide the results within a desirable time if traditional computing schemes are used.

29.5 UBIQUITOUS COMPUTING

Ubiquitous computing began in the Electronics and Imaging Laboratory of the Xerox Palo Alto Research Center (PARC) in the late 1980s. Mark Weiser (1952–1999) coined the phrase "ubiquitous computing" around 1988 during his tenure as a chief technologist at PARC [23]. Weiser wrote, in his paper entitled "The Computer for the 21st Century" [1]: "Specialized elements of hardware and software, connected by wires, radio waves and infrared, will be so ubiquitous that no one will notice their presence." In this paper and some subsequent papers by himself and with his colleagues at PARC, ubiquitous computing was defined and its details were sketched out. Weiser's 1991 paper starts with the following sentences: "The most profound technologies are those that disappear. They weave themselves into the fabric of everyday life until they are indistinguishable from it" [1].

Ubiquitous computing refers to the use of computers in everyday life, including smartphones and other mobile devices. It also refers to computers contained in commonplace objects such as cars and appliances. It implies computing where people are unaware of its presence. One of its features is that all these devices communicate with each other over wireless networks without any interaction by users. Ubiquitous computing is also called pervasive computing. All models of ubiquitous computing share a vision of small, inexpensive, robust network processing devices, distributed at all scales throughout everyday life, and in generally distinct connections [1, 23].

In particular, computers and networks are embedded within the complex social framework of daily activities, interplaying with the rest of our densely woven physical environment. Such an environment will become the truly computerized society of the 21st century. Weiser's idea of ubiquitous computing was influenced by many fields outside computer science, including philosophy, phenomenology, anthropology, psychology, and sociology.

Hiroshi Ishii (1956–) is a pioneer of the tangible user interface in the field of human-computer interaction. He founded the Tangible Media

Group when he joined the MIT Media Laboratory as a professor of media arts and sciences. Hiroshi Ishii and Brygg Ullmer wrote a paper entitled "Tangible Bits: Towards Seamless Interfaces between People, Bits and Atoms" [26]. Tangible bits allow users to grasp and manipulate bits in the center of users' attention by coupling the bits with everyday physical objects. When Weiser read the paper by Ishii and Ullmer, he noticed that the concept of ubiquitous computing and the concept of tangible bits are closely related. Weiser wrote an e-mail to Ishii, admiring the tangible bits research at MIT Media Laboratory, and further stated that this kind of work will characterize the technological landscape of the 21st century [24].

REFERENCES

1. M. Weiser, The Computer for the 21st Century, *Science America, Special Issue on Communications, Computers and Networks*, 265(3), 94–104, 1991.
2. A. S. Tanenbaum, *Computer Networks* (3rd ed.), Prentice-Hall, Upper Saddle River, NJ, 1996.
3. Wikipedia, Sputnik 1, http://en.wikipedia.org/wiki/Sputnik_1.
4. Wikipedia, Explorer 1, http://en.wikipedia.org/wiki/Explorer_1.
5. Wikipedia, Space Race, http://en.wikipedia.org/wiki/Space_Race.
6. Wikipedia, DARPA, http://en.wikipedia.org/wiki/DARPA.
7. Wikipedia, ENQUIRE, http://en.wikipedia.org/wiki/ENQUIRE.
8. Wikipedia, Packet Switching, http://en.wikipedia.org/wiki/Packet_switching.
9. Wikipedia, Rand Corporation, http://en.wikipedia.org/wiki/RAND_Corporation.
10. Paul Baran and Origins of the Internet, http://www.rand.org/history/baran.
11. Wikipedia, Paul Baran, http://en.wikipedia.org/wiki/Paul_Baran.
12. Wikipedia, ARPANET, http://en.wikipedia.org/wiki/ARPANET.
13. Wikipedia, CSNET, http://en.wikipedia.org/wiki/CSNET.
14. Wikipedia, J. C. R. Licklider, http://en.wikipedia.org/wiki/J._C._R._Licklider.
15. Wikipedia, Bob Kahn, http://en.wikipedia.org/wiki/Bob_Kahn.
16. Wikipedia, Vint Cerf, http://en.wikipedia.org/wiki/Vint_Cerf.
17. Wikipedia, World Wide Web, http://en.wikipedia.org/wiki/World_Wide_Web.
18. Wikipedia, Memex, http://en.wikipedia.org/wiki/Memex.
19. Wikipedia, Ted Nelson, http://en.wikipedia.org/wiki/Ted_Nelson.
20. Wikipedia, Tim Berners-Lee, http://en.wikipedia.org/wiki/Tim_Berners-Lee.
21. T. Berners-Lee and M. Fischetti, *Weaving the Web: The Original Design and Ultimate Destiny of the World Wide Web by Its Inventor*, Harper-Business, New York, 2000.
22. C. Kesselman and I. Foster (eds.), *The Grid 2: Blueprint for a New Computing Infrastructure* (2nd ed.), Morgan Kaufmann, San Francisco, 2003.
23. Wikipedia, Ubiquitous Computing, http://en.wikipedia.org/wiki/Ubiquitous_computing.

24. Wikipedia, Tangible User Interface, http://en.wikipedia.org/wiki/Tangible_user_interface.
25. L. Kleinrock, A Vision for the Internet, *ST Journal of Research*, 2(1), 4–5, 2005.
26. H. Ishii and B. Ullmer, Tangible Bits: Towards Seamless Interfaces between People, Bits and Atoms, in *Proceedings of the ACM SIGCHI Conference on Human Factors in Computing Systems*, 1997, pp. 234–241.
27. Wikipedia, Cloud Computing, http://en.wikipedia.org/wiki/Cloud_computing.
28. CG Technologies, Cloud Computing, http:w3.cgtechnologies.com/index.php?option=com_content&view=article&id=93.

CHAPTER 30

Public-Key Cryptography

30.1 THE SITUATION IN THE 1960s AND 1970s BEFORE THE PUBLIC KEYS

During the 1960s, the cost and performance of computers improved remarkably due to the development of electronics and semiconductor technology. In the early 1960s, IBM and other major computer manufacturers produced high-performance computers that used transistors and diodes in logic circuits in central processing units. Since then, computers have become more powerful and cheaper, and the computer market has been rapidly expanding ever since.

Before the 1960s, cryptosystems were mainly used by the government and various military and intelligence organizations. In the 1970s, private enterprises began to use cryptosystems for confidential data and communication security. For example, banks used computers to encrypt money transfers, and trade companies used them to encrypt transaction records. One of the primary problems is the issue of standardization of cryptosystems among companies. As more and more business companies used computers, the standardization of cryptosystems became an important issue. In the early 1970s, the U.S. government studied the needs for computer security and its standardization. In 1973, the U.S. standards body NBS (National Bureau of Standards, now called the National Institute of Standards and Technology) published the first request for a standard encryption algorithm. The second request by NBS was published in 1974; unfortunately, neither the first nor the second request turned out to be suitable. However, these requests by NBS eventually led to the adoption of the Data Encryption Standard (DES). DES was developed at IBM and became the most widely used cryptosystem by government organizations and business companies as well. DES was first published in the Federal Register in 1975, and adopted as a standard for unspecified applications in 1977. A complete description of DES was given in the Federal Information Processing Standards (FIPS) in 1977. By today's standards, DES is not considered completely safe.

A document is a sequence of sentences, which are strings themselves composed of letters or symbols from some finite alphabet (i.e., a finite set of symbols). Each letter or symbol can be expressed by a number. We may therefore consider that a document is simply a long sequence of numbers. In cryptography, such a sequence is usually divided into fixed-length groups of digits or bits, called *blocks*. Each block is transformed into a number by a block cipher encryption algorithm, giving us the *block cipher* cryptosystem (e.g., DES). Block ciphers are widely used for data security.

Since each block and each encrypted block may be considered a pair of numbers, an *encryption algorithm* is a function from integers to integers. For the same reason, a *decryption algorithm* is also a function from integers to integers. An encryption (decryption) key is used to specify an encryption (decryption) function, which is then implemented by an appropriate algorithm. When a sender (say, Alice) wishes to send a secret message (plaintext) to a receiver (say, Bob), Alice transforms the plaintext into its corresponding *ciphertext* by her encryption algorithm, and sends it to Bob. He receives the ciphertext, and decrypts it by applying his decryption algorithm.

For any cryptography appearing before the 1970s, the encryption key and decryption key are trivially related. Such cryptography is called a symmetric key, a shared key, a secret key, or a common key cryptography. DES is also a symmetric-key cryptosystem. For a symmetric-key cryptosystem, a sender (Alice) and a receiver (Bob) share the same secret key. The shared key should, obviously, be kept secret from an adversary or an eavesdropper (say, Eve). The number of possible key candidates is a crucial factor in determining the strength of a cryptosystem. If it is not large enough, a cryptanalyst will decipher the encrypted message by simply trying all possible keys. Therefore, the number of possible keys should be sufficiently large to ensure strong security of the cryptosystem. Otherwise, an eavesdropper could easily decipher the ciphertext with the help of a computer.

DES, which was adopted in 1977, encrypts a plaintext of a 64-bit string using a key. The key is a 56-bit string. The number of possible keys of DES is 2^{56}, or approximately 10^{17}. For such a big number of possible keys, it is practically impossible for anyone to correctly guess the secret key of a DES cryptosystem [5, 6]. However, in 1999, some analytical results demonstrated a theoretical weakness in the ciphers of DES. In 2002, DES was superseded by the Advanced Encryption Standard (AES). Unfortunately, both DES and AES are now considered insecure for many top-secret applications, because successful brute-force attacks are possible, although these systems are more than adequate for almost all commercial applications. No private company can afford to have powerful computers that can check every possible key within a reasonable amount of time. Of course, many foreign intelligence and military organizations may have sufficient resources and capabilities to break these codes.

The adoption of DES solved the problem of standardization for encrypting important and secret data. However, the problem of key distribution annoyed cryptographers. In secret-key systems, one of the major problems is the difficulty of secret-key exchange. This problem is known as *key distribution* [5]. It is defined to be a mechanism whereby one party chooses a secret key and then transmits it to another party or parties. In practical applications, the communication line itself may be insecure. We therefore need to protect it against potential adversaries. This problem was solved in the mid-1970s with the appearance of public-key cryptography [5, 6].

30.2 THE BIRTH OF PUBLIC-KEY CRYPTOGRAPHY

The ARPANET was created in 1969, and although it was still in its infancy in the early 1970s, some computer scientists and engineers predicted that with the advent of the Internet Age it would be indispensable for ordinary computer users to send their messages or data securely over the network. Whitfield Diffie (1944–) was one such scientist. He had been considering how such security could be guaranteed over the Internet.

In the best of all possible worlds, Internet users should be able to encrypt their messages so that no other network users other than legitimate receivers could decipher those encrypted messages. Such encryption might require the secure exchange or distribution of the encryption keys. Diffie was particularly interested in this problem (called the key distribution problem). He believed that techniques to overcome the key distribution problem would be very useful to construct a secure Internet world. However, solving this problem seemed very difficult.

In 1974, Diffie visited IBM's Thomas J. Watson Laboratory, where he gave a talk about the key distribution problem. Someone in the audience informed him that Martin Hellman (1945–) at Stanford University had been studying the same problem. Consequently, Diffie took a trip to California to meet Hellman, and began working with him on the key distribution problem. They were trying to find an efficient method for exchanging keys between a sender and a receiver on the Internet [5].

Diffie and Hellman thought that the encryption key could be public without decreasing the security of the encrypted message. This means that even if the eavesdropper knows the encryption key, he or she will still be unable to decrypt the ciphertext. Such an *asymmetric ciphersystem* is called a public-key cryptography. This idea was presented by Diffie and Hellman in the summer of 1975, and published in their joint paper [8].

How does public-key cryptography work? Suppose Alice wants to send a secret message to Bob. She puts it inside a box and closes the box with a padlocked key. Then she sends the padlocked box by post. Since she does not trust the postal employees, she delivers the key of the padlock to Bob by herself. She travels with the key to the location where Bob lives, and then she hands the key to Bob. From this story, we can appreciate the importance of key distribution for security. However, as each of the following three scenarios suggests, key distribution might be avoided without affecting security:

1. Assume that Bob sends an open padlock with Bob's identity (not including its key) to Alice by post. After Alice receives the open padlock, she puts her message in a box and locks the box with Bob's padlock. Then she sends the locked box to Bob by post. Since the postal employees do not have the key to Bob's padlock, the message can safely reach Bob. He opens the box with his own key and safely obtains the message from Alice. In this scenario, the message can be securely transferred from Alice to Bob.

2. Assume that Bob has a padlock and its key. Bob guards the key, but he manufactures thousands of replica padlocks. These replica padlocks without their keys are sold at post offices and supermarkets. When Alice wants to send a secret message to Bob, she buys a replica of Bob's padlock at a post office or a supermarket. She puts her secret message in a box and locks the box with the replica of Bob's padlock, and sends the locked box to Bob by post. Bob can obtain the secret message from Alice, since Bob has his own key for the padlock.

3. Assume again that Alice wants to send a secret message to Bob. She puts her message in a box and closes the box with her own pad-lock. Only Alice has a key for opening her padlock. Alice sends the locked box to Bob. When the box arrives, Bob adds his own padlock and sends the box back to Alice. Only Bob has a key for opening his padlock. When Alice receives the box, she removes her own padlock and sends the box back to Bob. Since Bob's padlock is still closed, the box safely arrives at Bob's location, and he can now open the box with his own key and read the secret message from Alice.

The idea of public-key cryptography is closely related to the ideas described in the scenarios above, and is also closely related to one-way functions. Suppose that given an argument value x, it is easy to compute $f(x)$, whereas it is intractable to compute x from $f(x)$. Such a function is called a *one-way function*. In the three scenarios above, the padlock (say, Bob's padlock) can be considered a one-way function $f(x)$ in the sense that computing the value $f(x)$ from x is analogous to the action for locking the padlock, whereas computing x from $f(x)$ is analogous to the action for unlocking the closed padlock. It is very hard for Eve, the eavesdropper, to unlock the closed padlock, since only Bob has the key of his padlock. Similarly, Alice's padlock can also be considered a one-way function. The key of a padlock corresponds to a trapdoor for computing x from $f(x)$. Such a one-way function with a trapdoor plays an essential role in realizing a public-key cryptosystem.

Diffie and Hellman focused their attention on one-way functions. A one-way function is relatively easy to compute for a given argument value, but the inverse computation is very difficult (intractable), especially if we do not have the secret information (called a trapdoor). Modular arith-metic is a rich area for one-way functions. For example, $f(x) \equiv r^x$ modulo q is easily computed, but in general, the computation from a given value (say, t) to x such that $t \equiv r^x$ modulo q is very hard if q is sufficiently large. Diffie and Hellman noticed that if a suitable one-way trapdoor function could be found, then it could be used as a public-key encryption function and the trapdoor could be used as a private key. Their idea was revolution-ary, appearing at a very late stage in the very long history of cryptogra-phy. Diffie and Hellman continued their research at Stanford University attempting to find a family of suitable one-way trapdoor functions, but they did not fully succeed in its discovery.

30.3 RSA CRYPTOGRAPHY

In 1977, Ron Linn Rivest (1947–), Adi Shamir (1952–), and Leonard Max Adleman (1945–) discovered a specific one-way trapdoor function. Rivest and Shamir are computer scientists, and Adleman is a mathematician, who was working on cryptography at MIT. In 1978, their public-key cryptography paper was published [9], and it is now referred to as the *RSA cryptosystem*, which stands for the first letters of Rivest, Shamir, and Adleman. Their one-way function is based on an amazingly simple number-theoretic idea, and yet it has successfully resisted all cryptanalytic attacks. The RSA (Rivest–Shamir–Adleman) cryptosystem has been most widely used in electronic commerce protocols, and it is believed to be secure if sufficiently long keys are used [1, 6]. In 2002, Rivest, Shamir, and Adleman received the Turing Award for their ingenious and practical contribution to the development of cryptography.

Suppose Bob wishes to set up an RSA system for anyone to send secret messages to him. He would then do the following:

1. Generate two large primes, p and q.

2. Compute $n = p \times q$ and $\varphi(n) = (p - 1)(q - 1)$, where $\varphi(n)$ is Euler's phi function that counts the number of positive integers less than n and relatively prime to n (see Chapter 20).

3. Randomly choose an integer b such that $1 < b < \varphi(n)$, and the greatest common divisor of b and $\varphi(n)$ is 1 (b and $\varphi(n)$ are relatively prime, i.e., gcd $(b, \varphi(n)) = 1$).

4. Compute $a \equiv b^{-1}$ mod $\varphi(n)$ using the extended Euclidean algorithm (i.e., $a \times b \equiv 1$ modulo $\varphi(n)$).

5. Publicize n and b, but keep the values p, q, and a secret. Bob's key is (n, p, q, a, b), where a is called his private key or secret key, and b is called his public key.

When Alice wishes to securely send her message x to Bob, she computes $y \equiv x^b$ modulo n using Bob's public key b, and then she sends the ciphertext y to Bob. When Bob receives the ciphertext, he computes $y^a \equiv (x^b)^a \equiv x^{ab} \equiv x$ modulo n by using his secret key a. In this way, Bob can easily read the message x from Alice. However, it is very hard for others to compute x from y, since $y \equiv x^b$ modulo n is a one-way trapdoor function and only

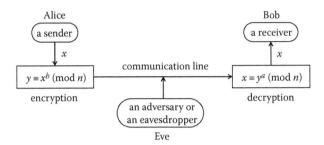

FIGURE 30.1   Communication on the RSA cryptosystem.

Bob has the secret key a. The fact that $x^{ab} \equiv x$ modulo n is based on the number-theoretic result, given in Theorem 30.1. Communication based on the RSA cryptosystem is shown in Figure 30.1.

Theorem 30.1

Let p and q be primes and $n = p \times q$. If $a \times b \equiv 1$ modulo $\varphi(n)$, then for any $1 \le x \le n - 1$, $x^{ab} \equiv x$ modulo n, where $\varphi(n)$ is Euler's phi function.   ■

Example 30.1

Let $n = 5 \times 11$. Then $\varphi(n) = \varphi(55) = (5 - 1)(11 - 1) = 40$. Suppose Bob chose $b = 3$; then Bob's public key is $n = 55$ and $b = 3$. Bob's secret key is $a \equiv b^{-1} \equiv 27$ modulo 40. If Alice wishes to send the message 13, she encrypts 13 to $13^3 \equiv 52$ modulo 55. Then Alice sends the ciphertext 52 to Bob. Bob receives 52 and decrypts it to $52^{27} \equiv 13$ modulo 55. Thus, Bob can read the original message 13 from Alice.

As shown in Example 30.1, modular exponentiation, i.e., computations of the form z^c modulo n, is necessary in communication by the RSA cryptosystem (x^b modulo n by Alice and y^a mod n by Bob). Computation z^c modulo n can be done using $c - 1$ modular multiplications. However, this naïve method is very inefficient if c is large. If c has k bits in its binary representation, the time complexity of this naïve method is exponentially large in k. We should use the square-and-multiply method to compute z^c modulo n, which reduces the number of modular multiplications to at most $2k$. Modular exponentiation by the square-and-multiply method is essential for the RSA cryptosystem since the computations by Alice and

Bob should be computationally feasible. For example, the exponential computation (52^{27} modulo 55 in Example 30.1) is carried out as follows:

$52^2 \equiv 9$ modulo 55

$52^4 \equiv (52^2)^2 \equiv 9^2 \equiv 26$ modulo 55

$52^8 \equiv (52^4)^2 \equiv 26^2 \equiv 16$ modulo 55

$52^{16} \equiv (52^8)^2 \equiv 16^2 \equiv 36$ modulo 55

$52^{27} \equiv 52^{16} \times 52^8 \times 52^2 \times 52 \equiv 36 \times 16 \times 9 \times 52 \equiv 13$ modulo 55

In order for the RSA cryptosystem to be secure, $n = p \times q$ must be large enough that factoring n will be computationally infeasible. Current factoring algorithms are able to factor numbers having up to 130 decimal digits. Hence, it is recommended that we pick p and q to be primes of about 100 digits. Then n will have about 200 digits. How can we find large primes of about 100 digits? Primary test algorithms can be used for this purpose.

The question of how to find large primes is especially important for a number of cryptographic protocols that use prime numbers. A naïve method for a primality test was proposed by an ancient Greek mathematician in an algorithm named the *sieve of Eratosthenes* (c. 240 BC). His method crosses out all multiples of primes and takes $O(n (\log n) \log \log n)$ bit operations. Another naïve method tries to divide a given number n by every number $m \le \sqrt{n}$. The time complexity of this method is $O(\sqrt{n}(\log n)^3)$. Therefore, these naïve methods are not useful at all for very large numbers.

Raymond E. Miller (1928–) and Michael O. Rabin (1935–), and Robert M. Solovay (1938–) and Volker Strassen (1931–) proposed probabilistic algorithms for efficiently determining if a given number is prime, in 1976 and 1977, respectively. The former is called the *Miller–Rabin* algorithm, and the latter is called the *Solovay–Strassen* algorithm. Both are classified as *yes-biased* Monte Carlo algorithms, and are quite useful in practice [6]. The running times of these algorithms are roughly $O((\log n)^3)$. A yes-biased Monte Carlo algorithm is a probabilistic algorithm for a decision problem in which a yes answer is always correct, but a no answer may be incorrect.

In 2002, Manindra Agrawal (1966–), Neeraj Kayal (1979–), and Nitin Saxena (1981–) proposed a deterministic algorithm for a primality test [10]. This is the first deterministic algorithm to test $O(\log n)$-digit numbers for their primality in time that has been proved to be polynomial in $\log n$. When they found the algorithm, Agrawal was a professor in computer science at the Indian Institute of Technology, Kanpur (IITK), and Kayal and Saxena were undergraduate students at IITK. Their algorithm attracted worldwide attention. In 2006, they received the Gödel Prize for their deterministic primality test algorithm. Named after the German mathematician Kurt Gödel, the Gödel Prize has been awarded jointly by the European Association for Theoretical Computer Science (EATCS) and the Association for Computing Machinery (ACM) since 1993 to outstanding papers in theoretical computer science.

The success of the Information Age owed much to the appearance of public-key cryptography. The simple encryption and decryption schemes of RSA cryptography have given a great advantage to the development of the Internet world. The public-key cryptosystems are indispensable, in particular, for secure data and message transfer, and for the secure operation of e-commerce, e-government, and worldwide interbank money transfers.

Although the RSA cryptosystem is the most well-known and widely used public-key cryptography, several other public-key cryptosystems have also been proposed. Of these cryptosystems, the following are of some significance [6]:

1. The *ElGamal* cryptosystem is based on the difficulty of the discrete logarithm problem for finite fields. It was first proposed by Taher ElGamal (1955–) in 1985.

2. The *McEliece* cryptosystem is based on algebraic coding theory. It was first proposed by Robert McEliece (1942–) in 1978.

3. The *elliptic curve* cryptosystem is based on the work in the domain of elliptic curves. The use of elliptic curves in cryptography was suggested independently by Neal Kobliz (1948–) and Victor S. Miller (1947–) in 1985.

30.4 DIGITAL SIGNATURES

A conventional handwritten signature attached to a document is used to specify the person responsible for it. A signature is often used in everyday situations such as writing a letter or a check, withdrawing money from the bank, or signing a contract. A *digital signature scheme* is a method of signing a document in electronic form, which can usually be transmitted over the Internet. When Bob receives an e-mail, how can he be sure that the message is really from Alice? The wicked Eve may write an e-mail and type Alice's name at the bottom of the message. How can we attach a signature to a document in an electronic form? How can we verify the authenticity of the signed document? Public-key cryptography can be applied to implement a digital signature scheme.

The following is an outline of a digital signature scheme by an RSA cryptosystem. Suppose that Alice wishes to send a secret message x to Bob. Let a and b be Alice's secret key and public key, respectively, and let c and d be Bob's secret key and public key, respectively. Remember that $a \times b \equiv 1$ modulo $\varphi(n)$ and $c \times d \equiv 1$ modulo $\varphi(n)$, and that for any message z, $z^{ab} \equiv z$ modulo n and $z^{cd} \equiv z$ modulo n. Alice's signature to the message x is defined as $y \equiv x^a$ modulo n. Alice encrypts pair (x, y) by Bob's public key d, and sends the encrypted pair to Bob. When Bob receives the encrypted pair, he decrypts it with his secret key c. Then he obtains (x, y). Bob can verify that the message is surely from Alice by checking whether y^b coincides with x. More formally, the signature scheme and the verification scheme by the RSA cryptosystem are described as follows:

1. Let $n = p \times q$, where p and q are primes.

2. The values n and b are public, and the values p, q, and a are secret, where $a \times b \equiv 1$ modulo $\varphi(n)$.

3. The signature of message x is defined as $y \equiv x^a$ modulo n (signature scheme).

4. (x, y) is true if and only if $x \equiv y^b$ modulo n (verification scheme).

Example 30.2

Let $(n, p, q, a, b) = (65, 5, 13, 11, 35)$ be Alice's key. Note that $65 = 5 \times 13$, $\varphi(65) = (5 - 1)(13 - 1) = 48$, and $11 \times 35 \equiv 1$ modulo 48. Suppose that Alice wishes to send the message 8 with her signature to Bob. Her signature is $8^{11} \equiv 57$ modulo 65. When Bob receives $(8, 57)$, he verifies that $57^{35} \equiv 8$ modulo 65. Then Bob is sure that the message 8 is truly from Alice. If the communication line from Alice to Bob is not secure, Alice may encrypt $(8, 57)$ by Bob's public key to prevent $(8, 57)$ from being known to eavesdroppers.

There are several digital signature schemes other than the RSA signature scheme. The *ElGamal signature scheme* is based on the difficulty of computing a discrete logarithm. It was first described by Taher ElGamal in a 1985 paper, "A Public-Key Cryptography and a Signature Scheme," in which he proposed the design of the ElGamal cryptography and the ElGamal signature scheme. The ElGamal signature scheme is nondeterministic. This means that there are many valid signatures for any given message. The verification algorithm must be able to accept any of these valid signatures. The ElGamal signature scheme must not be confused with ElGamal cryptography. All ElGamal signature schemes are designed specifically for the purpose of signature, as opposed to the RSA cryptosystem.

The original ElGamal signature scheme has been rarely used in practice. In 1989, C. P. Schnorr proposed a signature scheme that is a variant of the ElGamal signature scheme in which the signature size is significantly reduced. Another variant of the ElGamal signature scheme was developed in 1991 at the National Institute of Standards and Technology (NIST), incorporating some of the ideas in the *Schnorr signature scheme*. It was called the *digital signature scheme* (DSA) and adopted as a standard, specified in FIPS 186 in 1993. In 2000, the *elliptic curve digital signature algorithm* (ECDSA) was approved as a standard, specified in FIPS 186-2. It was a modification of the DSA to elliptic curves.

More information about cryptography can be found, for example, in [1–4, 6, 7].

30.5 ANOTHER STORY OF PUBLIC-KEY CRYPTOGRAPHY FROM ENGLAND

Since Diffie and Hellman published their paper "New Directions in Cryptography" in 1976 [8], they have been known as the first inventors of the concept of public-key cryptography. On the other hand, Rivest, Shamir, and Adleman have been known as the first inventors of the RSA cryptosystem. However, there is another story of an earlier invention involving the same idea of public-key cryptography [5, 11].

A few years earlier than Diffie and Hellman's invention of the concept of public-key cryptography, British mathematician James Henry Ellis (1924–1997) arrived at the same idea. Ellis worked in Communications-Electronics Security Group (CESG) of British Government Communications Headquarters (GCHQ). In 1970, Ellis wrote a paper, "The Possibility of Non-Secret Digital Encryption," in an internal report (CESG report). The nonsecret digital encryption proposed by Ellis is exactly the same idea as the public-key cryptography invented by Diffie and Hellman. In 1973, a young mathematician, Clifford Christopher Cocks (1950–), joined

GCHQ and was told about Ellis's nonsecret digital encryption. He thought that factoring an integer into prime numbers was a good candidate for a suitable one-way function. His idea was the same as RSA cryptography. Cocks's idea was 4 years prior to the invention of RSA cryptography by Rivest, Shamir, and Adleman.

Cocks and his colleague Malcolm John Williamson (1950–) worked together on the problem for realizing Ellis's idea. In 1973 and 1974, they wrote research reports about their invention of the nonsecret encryption algorithm, now known as the RSA encryption algorithm (C. C. Cocks, "A Note on Non-Secret Encryption," CESG Report, 1973; M. J. Williamson, "Non-Secret Encryption Using a Finite Field," CESG Report, 1974). These CESG reports were not publicized since they were treated as top-secret government information. Consequently, Ellis, Cocks, and Williamson's prior achievements remained unknown until 1997.

In 1987, Ellis wrote a paper, "The Story of Non-Secret Encryption." This paper had also been treated as an internal report within GCHQ. Although the invention of public-key encryption at GCHQ had not been publicized by 1997, GCHQ in UK and the National Security Agency (NSA) in the United States knew about the work of Ellis, Cocks, and Williamson. In December 1997, Cocks delivered a public talk that contained the history of GCHQ's research on public-key cryptography. Since then, Ellis, Cocks, and Williamson's contributions to the concept and realization of public-key cryptography have been acknowledged [5, 11].

REFERENCES

1. A. Salomaa, *Public-Key Cryptography*, Springer-Verlag, Berlin, 1990.
2. O. Goldreich, *Modern Cryptography, Probabilistic Proofs and Pseudo-Randomness*, Springer-Verlag, Berlin, 1999.
3. O. Goldreich, *Foundations of Cryptography: Basic Tools*, Cambridge University Press, Cambridge, UK, 2001.
4. O. Goldreich, *Foundations of Cryptography: Basic Applications* (vol. II), Cambridge University Press, Cambridge, UK, 2004.
5. S. Singh, *The Code Book: The Science of Secrecy from Ancient Egypt to Quantum Cryptography*, Anchor Books, New York, 1999.
6. D. R. Stinson, *Cryptography: Theory and Practice* (3rd ed.), Chapman & Hall/CRC, New York, 2006.
7. J. von zur Gathen and J. Gerhard, *Modern Computer Algebra*, Cambridge University Press, Cambridge, UK, 1999.
8. W. Diffie and M. Hellman, New Directions in Cryptography, *IEEE Transactions on Information Theory*, IT-22, 644–654, 1976.

9. R. Rivest, A. Shamir, and L. Adleman, A Method for Obtaining Digital Signatures and Public-Key Cryptosystems, *Communications of the ACM*, 21, 120–126, 1978.

10. M. Agrawal, N. Kayal, and N. Saxena, PRIMES is in P, *Annals of Mathematics*, 160, 781–793, 2004.

11. Wikipedia, James H. Ellis, http://en.wikipedia.org/wiki/James_H._Ellis.

CHAPTER 31

Quantum Computing

Quantum computing, based on the manipulation of the smallest atomic particles, brings together aspects of quantum physics, mathematics, and computer science. Using this three-way grouping provides a new approach to computation that is very different from the now ubiquitous digital approach. Quantum computing, although showing great promise, is still unproven, and it is not yet clear how useful and powerful it will ultimately be.

31.1 THE BASICS OF QUANTUM COMPUTING

The development of radio techniques and the improvement of other technical aids to study physical phenomena led at the end of the 19th century to the discovery of electrons, x-rays, and radioactivity. However, classical physics was just not able to explain the properties of atomic and subatomic particles. A study of conditions of equilibrium between matter and electromagnetic radiation by Max Planck (1858–1947) in 1900 and of photoradiation phenomena by Albert Einstein (1879–1955) led to the conclusion that electromagnetic radiation possessed both a wave character and a discrete particle character. It was the start of quantum physics.

Quantum physics joined the mainstream in the 1920s and 1930s with the general acceptance of the theories of Max Planck, Albert Einstein, Niels Bohr (1885–1962), Erwin Schrödinger (1887–1961), Werner Heisenberg (1901–1976), Paul Dirac (1902–1984), and many other established physicists. Quantum computing is based on quantum physics, with all of its special behaviors and unusual limits. Therefore, quantum computing deals with the behaviors of atomic and subatomic particles. These behaviors are irreducibly random and the measurement of particle characteristics simultaneously, such as position and momentum, to an arbitrary precision is impossible. That is, physical phenomena of small particles do not agree often with our classical intuition. The unusual behavior results from features of quantum mechanics called *superposition* and *interference*. In the early 1980s, Richard Feynman (1918–1988) noted that there seemed to be fundamental difficulties in simulating quantum mechanical systems on digital computers, and further suggested that having computers based on the principles of quantum mechanics would overcome those difficulties. Devices that perform quantum information processing are known as quantum computers.

Whereas the lowest-level unit in common digital computation is the binary digit, the bit, the lowest-level unit in quantum computation is the quantum binary digit, the qubit [1]. As described in [1], about 10^3 atoms are typically used to store 1 bit of information [2, 3]. Within a quantum environment, one subatomic particle can encode one, two, or even more qubits. The physical size of qubits is much smaller than bits; nevertheless, their behavior is much more complex. Qubits can be put into a superpositional state. When they are in superposition, the qubit simultaneously has multiple values. Qubits can also become entangled, which is explained below. In addition, qubits can encounter decoherence; here, the qubit unexpectedly changes state and may lose some of its special behaviors, such as superposition.

The behavior of an individual qubit depends on its current state, which is the result of its history. Qubits normally start out in one of two states. These states are frequently described using the "ket" style of notation, using |0> and |1>, respectively, for binary states 0 and 1. However, once a qubit is put into superposition state, it is not bounded to only one of these two values. The superposition state is not readily observable because when a qubit in a state of superposition is observed (measured), it immediately collapses to one of the two binary states: |0> or |1>.

The linear algebra notation in quantum computation may not be familiar to a student of mathematics or computer science. The notation was invented by Paul Dirac and is known as Dirac notation. This notation is used often in quantum mechanics. The Bloch sphere is also often used to represent the state of a single, unentangled qubit, and it makes it possible to view the state of the qubit in a graphical manner. In Figure 31.1, a qubit showing the value |1> is depicted, which is represented by –1. Qubit states are often represented with three probability amplitudes, along the x-axis, y-axis, and z-axis, respectively. In this figure, the x- and y-values are zero and the z-value is –1; therefore, the arrow is pointing downwards. These

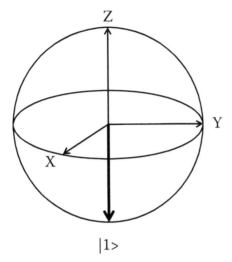

|1>

FIGURE 31.1   Bloch sphere showing a value of |1>.

probability amplitudes are expressed via complex numbers, and these amplitudes can be both positive and negative. In this particular figure, the three values are all real numbers. With quantum behaviors, adding two probabilities together may result in a reduced probability.

Qubits may become entangled. Entanglement is one of the most unusual behaviors in quantum theory. Once two qubits are entangled, neither one can be described without full mention of the other. Even if the two qubits are physically separated after entanglement, impacting one qubit impacts the other. Entanglement was described by Einstein as "spooky actions at a distance" (*spukhafte Fernwirkungen*) [10]. Entanglement is a basic feature of quantum computation and quantum communications.

As previously mentioned, measurement may change the state of a qubit. Therefore, it is not generally possible to copy a qubit by measuring its value and providing another qubit with a matching value. How can qubits be manipulated and processed if intermediate values cannot be measured or observed? Qubit manipulations are done via transformations of the qubit's probability amplitudes. With quantum programming, these probability amplitudes are manipulated so that, upon measurement, the desired values are observed in the qubits. Quantum processing also requires that all quantum actions be reversible (Landauer's principle). The result of this is that all gates and circuits used with quantum computing have the same number of inputs and outputs. This implies that information cannot be lost or erased by these quantum transformations.

31.2 QUANTUM COMPUTATION LOGIC AND GATES

Just as digital computation utilizes logic gates, quantum computation also utilizes gates. The quantum gates are very different from the digital gates, partially due to their need for reversibility and the same number of inputs and outputs. With quantum computing, some of the common gates are the Hadamard gate (H gate for short), the controlled-NOT gate, the Toffoli gate, and the Pauli-X, Pauli-Y, and Pauli-Z gates. The behavior of quantum gates is often expressed in the form of a unitary matrix. The remainder of this section provides more information about the H gate and the controlled-NOT gate.

The H gate has one input and one output. This gate is commonly used to put a qubit into a state of superposition. Its behavior can be described by this unitary matrix: $\frac{1}{\sqrt{2}}\begin{bmatrix} 1 & 1 \\ 1 & -1 \end{bmatrix}$. When an H gate is applied two times in sequence, it returns the original value.

The controlled-NOT gate is more interesting than the H gate. It has two inputs and two outputs. Note that the input qubits are usually mutually entangled by this gate. The basic processing done is an exclusive OR operation (XOR) of the inputs, with one input being passed directly through to the output; this is called the control input. And the other output contains the results of the XOR operation. If the second, noncontrol, input is held to |0>, then this gate functions as a NOT operation. The unitary matrix for the controlled-NOT operation is $\begin{bmatrix} 1 & 0 & 0 & 0 \\ 0 & 1 & 0 & 0 \\ 0 & 0 & 0 & 1 \\ 0 & 0 & 1 & 0 \end{bmatrix}$.

The Toffoli gate is a three-input, three-output gate that is similar to the controlled-NOT, excepting that it does the XOR with the AND of two control inputs. The Pauli-X, Pauli-Y, and Pauli-Z gates are rotations about the respective x-, y-, and z-axes. The Paul-X gate can function as a NOT operation.

31.3 FAMOUS QUANTUM ALGORITHMS

In 1985, David Deutsch (1953–) attempted to define computational devices that would be capable of efficiently simulating an arbitrary physical system. Deutsch was naturally led to consider computing devices based upon the principles of quantum mechanics. These devices led to the modern concept of a quantum computer.

Three quantum computing milestone algorithms are now discussed, in order of increasing algorithm complexity. The basic steps in these algorithms tend to be: establish the qubits in classic states, put some of the qubits into superposition, apply a set of unitary operations (transformations), and then measure the values of the desired subset of the qubits. Often the results of the quantum algorithms are frequently further processed by digital computation, sometimes including numerous nonquantum algorithmic operations.

31.3.1 Deutsch's Algorithm (1989) [1]

This algorithm solves a problem with a binary result. The problem can be expressed as making a yes/no financial investment decision that is based on the result of two long-running calculations, each returning a binary result. When equal values are returned, an action is taken. Quantum computing allows these two long-running calculations to occur in parallel. This problem can also be expressed mathematically.

The two results of the function f: $\{0, 1\} \rightarrow \{0, 1\}$ can be *constant* or *balanced*. They are constant if $f(0) = f(1)$ and balanced if $f(0) \neq f(1)$. With classic digital computation, $f(x)$ needs two calculations, one calculation for $f(0)$ and the other for $f(1)$. With superposition, parallel calculation of both $f(0)$ and $f(1)$ is done in the time needed for just a single calculation. The quantum circuit to compare the results is a *Hadamard* gate providing the needed superposition, the long-running calculation done as unitary transformations, and then a controlled-NOT gate to compare the two results. The controlled-NOT, performed in parallel, brings $|x, y>$ to $|x, y \oplus f(x)>$, where $\oplus$ is XOR. The two results are then put through Hadamard gates. In simplified terms, when the $\oplus$ result is 0, the results are constant, and when 1, balanced. It is interesting to note that the direct results of the two time-consuming calculations are never observed; what is observed is the constant or balanced relationship between them. Figure 31.2 shows a version of Deutsch's algorithm as a (simplified) quantum circuit.

31.3.2 Grover's Search Algorithm (1995) [1, 4, 5]

Further evidence for the power of quantum computers came in 1995 when Lov Grover (1961–) showed that the problem of concluding a search through some unstructured search space could be sped up on a quantum computer. Grover's search algorithm can be viewed as a solution to a number of different problems, including looking up a value in a database,

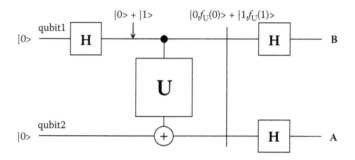

FIGURE 31.2 Deutsch's algorithm in circuit form.

finding a needle in a haystack, and inverting a function. The notion is that an unordered list of values is searched for a particular value. Utilizing classic computation, on average, $O(n)$ time is needed to find a particular value. Utilizing Grover's quantum algorithm, $O(\sqrt{n})$ time is needed, where n is the database size. This provides a quadratic speed-up and can be substantial when n is large.

From the inverting a function point of view, the function f: $\{0, 1\}^n \rightarrow \{0, 1\}$ is evaluated in search of the unknown **value**, which is n bits wide, such that **value** = x if $f(x)$ is 1, x varying over the domain and only one result of f will provide a 1. This quantum algorithm operates by creating all possible input values, via superposition, and then the one input value that generates the desired output value is *transformed* such that when the qubits are measured, this input value is present in the qubits. Finding the desired input value at measurement requires the use of *phase inversion* and *inversion about the average* transformations, which are done in multiple steps. Unfortunately, in a small number of cases the correct value will not be measured and the algorithm fails.

Grover's algorithm shows the solution to a real-world problem using a quantum algorithm, and this algorithm provides a substantial improvement in performance.

31.3.3 Shor's Factoring Algorithm (1994) [1, 6]

Before discussing this important factoring accomplishment, we require a small detour, away from algorithms. In 1981, Richard P. Feynman brought the notion of quantum computing into full view when he noted in his famous speech, "Simulating Physics with Computers," that quantum computing might provide answers more quickly, relative to digital computers, when performing quantum operations [7]. Digital computers often have performance and storage capacity issues when handling large quantum simulations. This is due to the explosion of terms that occurs when the tensor product of a large number of quantum particles is formed. Feynman proposed this high-value quantum computing use and helped provide quantum computing the beginnings of legitimacy.

The algorithm by Peter W. Shor (1959–) gave quantum computing additional legitimacy, as it provides a fast method to factor numbers. Factoring large numbers is a difficult problem, but is not considered a NP-complete problem. It is considered to be between NP and P. It is exactly what gives many security algorithms their power [8]. If a practical way were found to rapidly factor any large number, then the commonly used security approaches like RSA (See Chapter 30) and SSL (Secure Socket Layer) would no longer be secure. Shor's algorithm outlines a workable approach to factoring large numbers, but there are practical realities, discussed below, that prevent the algorithm from being really useful.

This algorithm is complex and uses both classical computation and quantum computation. The basic approach is to find a root of an odd composite number. Once this root is known, then with additional steps, the odd number can be quickly factored. The approach is rooted in the knowledge that the factoring problem can be reduced to finding the period of a repeating function. The quantum part of this algorithm involves finding the period of the function. This determination is done by using a superposition and then applying a function on all of the superposition values, looking for the point where the repeating begins. A quantum *Fourier transform* is applied to isolate the period. From this stage, classical computation is done to find the actual factors of the composite number.

Shor's algorithm makes it clear that quantum computing can solve real-world problems, and, along with Feynman's earlier work, makes it clear that quantum computing has significant potential.

31.4 DIFFICULTIES AND LIMITS OF QUANTUM COMPUTING

So, why is quantum computing not in the mainstream of computing? A hint of why comes from the fact that the IBM researchers who first implemented Shor's algorithm in 2001 using nuclear magnetic resonance (NMR) actually factored 15 into 3 times 5 [9]. In practical terms, this is hardly a great achievement. The researchers were limited by the number of qubits they could manipulate without the qubits suffering decoherence, due to qubit's close proximity and unwanted interactions.

It turns out that quantum computing hardware is not easily constructed, nor is it easily scaled. As of this writing, some of the common approaches used to construct qubits are based on *ion traps*, *linear optics*, *NMR*, and *superconductors*. However, all of these technologies have problems with decoherence, and a register of more than seven or eight qubits is considered large. Since 2^8 equals only 256, this is a very significant limit in terms of the power of quantum computation. R. Van Meter and Clare Horsman deftly summarize the status of quantum computing hardware when they pose the question, "When will a quantum computer do science, rather than be science?" [11].

In addition to the quantum computing hardware issues and limits, software for quantum computing is not easily produced. Quantum algorithms, often implemented as quantum circuits, are not created with the *mathematical*, *if/then*, and *loop* operations common to digital algorithms, computation, and programming. Instead, quantum algorithms are created using transformations, and evolutions, of probability amplitudes. In mathematical terms, this amounts to programming strictly via the manipulations provided by the multiplication of (large) unitary matrices. This is not an easy way to program.

31.5 CLOSING SUMMARY

Using atomic or subatomic particles, quantum computing draws together quantum physics, mathematics, and computer science. Using this approach to computation is very different from the now ubiquitous digital approach. Successful quantum algorithms have been created, and the quantum logic gates needed for the construction of large quantum circuits do exist. However, due to various significant difficulties and limits, it is not yet clear how useful and powerful quantum computing will ultimately prove to be.

REFERENCES

1. N. S. Yanofsky and M. A. Mannucci, *Quantum Computing for Computer Scientists*, Cambridge University Press, New York, 2008.
2. M. Hayward, Quantum Computing and Shor's Algorithm, April 26, 2008, url: alumni.imsa.edu/~matth/quant/299/paper/.
3. C. Williams, *Explorations in Quantum Computing* (2nd ed.), Springer-Verlag, New York, 2011.
4. C. Lavor, L. Manssur, and R. Portugal, Grover's Algorithm: Quantum Database Search, ARXIV, quant. ph. 1079L, 2003.
5. Wikipedia, Grover's algorithm, http://en.wikipedia.org/wiki/Grover's_algorithm.
6. Wikipedia, Shor's algorithm, http://en.wikipedia.org/wiki/Shor's_algorithm.

7. R. Feynman, Simulating Physics with Computers, *International Journal of Theoretical Physics*, 21(6/7), 467–488, 1982.
8. S. Aaronson, The Limits of Quantum, *Scientific American*, 298, 62–69, 2008.
9. L. M. K. Vandersypen, M. Steffen, et al., Experimental Realization of Shor's Quantum Factoring Algorithm Using Nuclear Magnetic Resonance, *Nature*, 414(6866), 883–887, 2001, doi: 10.1038/414883a.
10. Letter from Einstein to Max Born, March 3, 1947, in *The Born-Einstein Letters; Correspondence between Albert Einstein and Max and Hedwig Born from 1916 to 1955*, Walker, New York, 1971 (cited in M. P. Hobson, et al., *Quantum Entanglement and Communication Complexity*, 1998, pp. 1–13, CiteSeerX: 10.1.1.20.832).
11. R. Van Meter and C. Horsman, A Blueprint for Building a Quantum Computer, *Communications of the ACM*, 56(10), 84–93, 2013.

Index

A

Abacus, 59
  Chinese, 8–9, 64–65
  earliest systems, 59–61
  Greek and Roman, 61–63
  Japanese, 65–66
  not true calculating machine, 85
  sand, 9, 59
Abbreviated Computer Instructions, 157
ABC (Atanasoff-Berry Computer), 149–150
Abel, Niels Henrik, 102–103
Abramson, N., 201
ACE (Automatic Computing Engine), 145,
  156–157
Acrophonic system, 61
Ada, 234
Adleman, Leonard Max, 300
Advanced Encryption Standard (AES), 297
Advanced Research Projects Agency
  (ARPA), 273, 283, 285
Aeolipile, 205
AES, 297
African Paleolithic evidence, 3
Agrawal, Manindra, 302
Ahmes, 24
AIBO, 224
Aiken, Howard, 150
Akhmim wooden tablets, 16
Alberti, Leon Battista, 139, 143
Algebra
  Boolean, 121–128
  Diophantus and *Arithmetica*, 43–49
  "Father of," *See* Al-Khwarizmi
  fundamental theorem of, 112

Algebraic equations, 73, 97
  cubic, 100–101
  numerical solutions, 164–170
  quadratic, 99–100, *See also* Quadratic
    equations
  quartic and quintic, 101–105
ALGOL, 234, 236–237, 256
Algorithms, 245–247
  al-Khwarizmi and, 100
  Bellman-Ford, 263
  cardinality-based analyses, 119–120
  complexity of, 246–253
  Dijkstra's, 262–263, 281
  efficiency of, 250, 256
  Euclidean, 37–39, 71, 134, 245
  Floyd's, 263
  formalisms, 135
  Gödel's incompleteness theorem and,
    130
  graph, 260–264
  Monte Carlo, 266–268, 302
  parallel, 275–279
  primality tests, 267–268
  programming languages and, 246,
    See also Computer algorithmic
    design
  quantum, 312–315
  randomized, 265–268
  recursive, 257
  sorting and searching, 255–258, 265
  Turing machines, 133, 135
Al-Jazari, 205–206
Al-Kashi, Jamshid, 77–78
Al-Khwarizmi, 12, 73, 100, 245
Al-Kindi, 138–139
Allen, Frances E., 272

Altman, Mieczyslaw, 169–170
Amazon, 290
Analytical engine, 92, 93, 220
AND gates, 125
AND operations, 121–122
Andrica, Dorin, 33
A Programming Language (APL), 259–260
Arab cryptanalysts, 137–139
Arabic mathematics, *See also*
      al-Khwarizmi
  Euclid's *Elements* and, 40
  use of Indian numeral system, 12–13
Arabic numerals, *See* Hindu-Arabic
    numerals
Archimedes, 79, 162
Argand, Jean-Robert, 111
Aristotle, 35, 124
Arithmetica (Diophantus), 43–49
Arithmetic operations, 13, *See also*
    Division; Multiplication
  arithmetic expressions, 210
  time complexity of, 247–248
Arora, Sanjeev, 253
ARPANET, 286–287, 297
Arrays, 258
Artificial intelligence (AI), 217
  areas of, 227–229
  chess programs, 217, 220, 227
  Dartmouth conferences, 221, 224, 226
  defining, 218–219
  expert systems, 217–218, 219, 222, 223,
    228
  LISP and, 222, 223, 234
  origin of term, 221
  pioneers, 224–227
  timeline, 219–224
  Turing test, 133
Artificial Intelligence Laboratory at MIT,
    222
Aryabhata, 12, 177
ASIMO, 224
Asimov, Isaac, 221
Assilian, Seto, 222
Asymmetric cipher system, 298
Asynchronous model, 275, 279
Atanasoff, John Vincent, 149
Atanasoff-Berry Computer (ABC), 149–150

Automata, 205–206
  chess player, 220
  generative grammars and, 214
  I/O automaton, 279
  Turing machines and computing
    models, 206–211
Automatic Computing Engine (ACE), 145,
    156–157

B

Babai, Laszlo, 267
Babbage, Charles, 90, 91–92, 93, 141, 220
Babylonians, 7, 8, 11, 60, 162
Backus, John, 232, 237
Backus-Naur Form (BNF), 237
Ballistic calculation, 153–154, 183
Barak, Boaz, 253
Baran, Paul, 284
Bar-Hillel, Y., 214
Base-12 (duodecimal) number system,
    10, 63
Base-60 (sexagesimal) number system, 7,
    8, 10, 60
BASIC, 238
BCH codes, 202
Beer, Stafford, 182
Bellman, Richard, 263
Bellman-Ford algorithm, 263
Berners-Lee, Tim, 288–289
Bernoulli numbers, 93
Bernstein, Alex, 226
Berry, Clifford, 149
Beyer, Johann Hartmann, 78
BINAC, 154
Binary number system, 151
Binary search method, 257
Binet, J. P. M., 74
Binet's formula, 74
Bit, 184, 226
Bloch sphere, 310
Block ciphers, 296–297
Block codes, 194–196
Blocks Microworld Project, 222
Blum, Manuel, 249
Blum complexity algorithms, 249
Bohr, Niels, 309
Bomba, 144

Bombe, 145, 151
Bombelli, Rafael, 47
Bonfils, Immanuel, 77
Book of Calculation (*Liber Abaci*, Fibonacci),
    69–72, 81, 167, 177–178
Boole, George, 121, 151
Boolean algebras and logic, 121–128, 151
    Boolean functions, 125–128
    NP-complete problems, 252
    switching theory and, 127
Bose, R. C., 201
BPP complexity class, 268
Brahmagupta, 47, 100, 177
Briggs, Henry, 80–81
Brun, Viggo, 32
Bubblesort, 256
Buchholz, Werner, 272
Bullynck, Maarten, 178
Burgi, Jobst, 78, 81
Burks, A. R., 149
Bush, Vannevar, 126, 287

C

C, 234
C#, 238, 242–243
C++, 240–241
Caesar, Julius, 54–55
Caesar cipher, 54–55, 137
Cailliau, Robert, 288
Cairo wooden tablets, 16
"Calculating clock" (*Rechen Uhr*), 86–87
Calculating machines, 85, *See also*
    Electronic computers
    Babbage's mechanical computers, 91–92
    Hollerith and the Tabulator, 93–95
    Jacquard loom, 89–90, 219
    Pascaline, 87–88, 219
    Schickard's "calculating clock," 86–87
    Stepped Reckoner, 88–89
Calculi, 5, 59–60
Calendars, 2, 3, 13
Cantor, Georg, 108, 109, 115–119, 130
Capek, Josef, 220
Capek, Karel, 220
Cardano, Girolamo, 101, 111
Cardinality, 115–120
Cave art, 1–2

Cayley, Arthur, 169
Census data processing, 94–95
Cerf, Vinton, 286
Chebyshev, Pafnuty, 28
Checker program, 222, 226
Chess programs, 217, 220, 226, 227
China
    abacus, 8–9, 64–65
    algebraic equations, 98
    decimal fractions, 77
    secret messages, 52
    use of zero, 12
Chinese remainder theorem, 175–178
Chomsky, Avram Noam, 211–214
Church, Alonzo, 134–135, 155, 245
Church-Turing thesis, 134–136
Clay tokens and tablets, 5, 59
Client-server model, 289
Clock arithmetic, 173–175
Cloud computing, 289–291
COBOL, 237–238
Cocke, John, 272, 273
Cocks, Clifford Christopher, 305–306
Coding, *See also* Cryptography;
        Programming languages
    block, 194–196
    cyclic, 201–202
    Golay, 200, 202
    Hamming, 196–198, 199, 201
    Huffman, 188–189
    linear, 198–202
    Morse, 189–190, 191t
    parity check, 193–194, 200
    random error checking, 201–202
    Reed-Muller, 201
    Shannon-Fano, 186–188
Cohen, Paul, 118
Colmerauer, Alain, 222, 239
Colossus, 146, 151–153, 220
Commensurable and incommensurable
    numbers, 39
Common logarithm, 79
Commonsense knowledge and reasoning
    AI, 229
Complexity, *See* Computational complexity
Complex numbers, 110–112, 118
Complex-valued functions, 112–113

Computability, 129
    Church-Turing thesis, 134–136
    Gödel's incompleteness theorem,
        129–130
    total functions, 130–131
    Turing machines, 131–134
Computational complexity, 246–253
    addition and multiplication
        algorithms, 247–248
    complexity classes, 249–250, 268
    efficient computation, 250
    nondeterministic, 251
    NP-complete problems, 251–253
    parallel algorithms, 278
Computer algorithmic design, 246, 255
    data structures, 258–260
    dynamic programming, 263–264
    graph algorithms, 260–264
    randomized algorithms, 265–268
    sorting and searching, 255–258
Computer networks, 283
    ARPANET and CSNET, 284–287
    cloud and grid computing, 289–291
    packet switching, 283–284
    ubiquitous computing, 291–292
    World Wide Web and hypertext,
        287–289
Computer programming, *See*
        Programming languages
Computers, *See* Electronic computers
Computer vision, 228
Computing models, Turing machines as,
        206–211
Concatenation, 209–210
Congruence, 30
Context-free grammars and languages
        (CFGs and CFLs), 213, 214
Context-sensitive grammars and
        languages (CSGs and CSLs),
        213, 214
Continuum hypothesis, 118
Controlled-NOT gate, 312, 313
Cook, Stephen Arthur, 251–253, 278
Copernicus, Nicolaus, 40
Cotes, Roger, 112
Couffignal, Louis, 182–183
Counting, history of, 1
    earliest abaci, 59–61

paleolithic art, 1–2
tally sticks and representational
        symbols, 2–4
use of fingers, 2, 10
use of pebbles, 4–5, 61
use of tokens, 5, 59
Cray computers, 273–274
Crelle, August, 103
Cro-Magnon, 1–3
Cryptography, 52, *See also* Public-key
        cryptography
    ancient methods, 52–56, 137
    Arab cryptanalysts, 137–139
    block ciphers, 296–297
    cryptographic machines, 143–146,
        151–152, *See also* Enigma
        machine
    DES and AES, 296–297
    homophonic substitution, 141–143
    key distribution problem, 297–299
    monoalphabetic substitution, 53–55,
        138–139
    1960s and 1970s systems, 295–297
    polyalphabetic substitution, 139–141
    Shannon's work, 183
    standardization issues, 295–296
CSNET, 286–287
Cuneiform symbols, 7, 60
Cybernetics, 181–183
Cyclic codes, 201–202

D

d'Alembert, Jean Le Rond, 112
Darius Vase, 62
DARPA Grand Challenge, 224
Dartmouth conferences, 221, 224, 226
Data Encryption Standard (DES), 296–297
Data mining, 228
Data structures, 258–260
Davies, Donald, 284
Da Vinci, Leonardo, 206
Decibel, 82
Decimal fractions, 77–78
Decimal number system, 9–10, 12–13, 100
    abacus and, 60
    Fibonacci's *Liber Abaci* and, 70–71
    rational numbers and, 18

Dedekind, Julius, 108–109
Dedekind cut, 109
Defense Advanced Research Projects
      Agency (DARPA), 224, 283
Delphi, 242
De Morgan, Augustus, 122, 123–124
DENDRAL, 222
DES, 296–297
Descartes, Rene, 108
Deterministic computations, 250
Deutsch, David, 312
Deutsch's algorithm, 313
Dickmanns, Ernst, 223
Diesel Electric Locomotive
      Troubleshooting Aid, 223
Difference engine, 91–92
Differential analyzer, 126–127
Differential equations, 171
Diffie, Whitfield, 297–299, 305
Digit, 2
Digital signatures, 303–305
Dijkstra, Edsger Wybe, 262–263, 280
Dijkstra's algorithm, 262–263, 281
Diophantine equations, 38, 47, 48
Diophantus of Alexandria, 43–49
Dirac notation, 310
Dirac, Paul, 309, 310
Distributed computing, 279–281
   cloud and grid computing, 290–291
Division
   algorithm in *Liber Abaci*, 71
   ancient methods, 15
   Stepped Reckoner and, 88
Doolittle's algorithm, 166
Dunwell, Stephen W., 272
Duodecimal (base-12) number system,
      10, 63
Dynamic programming, 263–264

E

e, 108
   Euler's formula, 112–113
Earthquake scale, 82
Eckert, John Presper, 153–154
Edmonds, Dean, 221
EDSAC, 158
Efficient algorithms, 250, 256

Egypt, ancient, 15
   algebraic equations, 98
   number representation, 7, 10
   prime numbers and, 24
   use of fractions, 16–17
Eigenvalues and eigenvectors, 171
Einstein, Albert, 29, 164, 309
Elastic Compute Cloud (EC2), 290
Electronic computers, 149, *See also*
      Parallel computing
   ABC, 149–150
   ACE, 145, 156–157
   AI and, 217
   Colossus, 151–153, 220
   EDSAC, 158
   ENIAC, 153–155, 170, 220
   LARC, 272
   Mark I, 157–158
   numerical analysis and, 170
   programming, *See* Programming
      languages
   SEAC, 158
   supercomputers, 272, 273–275
   SWAC, 158
   von Neumann architecture, 155–156,
      170, 220, 275
   Whirlwind I, 158
   Z3, 150–151
Electronic Delay Storage Automatic
      Calculator (EDSAC), 158
Electronic Discrete Variable Automatic
      Computer (EDVAC), 155
Electronic Numerical Integrator and
      Calculator (ENIAC), 153–155,
      170, 220
ElGamal, Taher, 303, 305
ElGamal cryptosystem, 303
ElGamal signature scheme, 304–305
Elliptic curve cryptosystem, 303
Elliptic curve digital signature algorithm
      (ECDSA), 305
Ellis, James Henry, 305–306
ENIAC, 153–155, 170, 220
Enigma machine, 143–144, 151, 220
Entanglement, 311
Entropy, 184–186
Eponymous paradox, 118
Eratosthenes, 26

Erdös, Paul, 47
Erdös-Straus conjecture, 47
Error detecting and correcting codes
   block codes, 194–196
   cyclic codes, 201–202
   Golay code, 200, 202
   Hamming codes, 196–198, 199, 201
   linear codes, 198–202
   parity check, 193–194
   parity check matrix, 200
   random error checking, 201–202
   Reed-Muller codes, 201
   Shannon's information theory and, 194
   space exploration applications, 202
Etruscan Cameo, 62
Euclid, 24–25
Euclidean algorithm, 37–39, 71, 134, 245
Euclid's *Elements*, 24–26, 35–40, 70, 245
Eudoxus, 35, 38
Euler, Leonhard Paul, 28, 74, 102, 112, 162, 176, 179, 261
Euler's formula, 112–113
Euler's phi function, 179
Everett, Robert, 158
Expert systems, 217–218, 219, 222, 223, 228
Explorer 1, 283
Exponential function, 79
Extrapolation, 171

F

Factoring algorithm, 314–315
False position method, 167
Fano, Robert Mario, 186
Fast Fourier transforms, 248
Feedback loops, 182–183
Feigenbaum, Edward, 222
Fermat, Pierre de, 27, 47, 87, 178
Fermat primes, 30
Fermat's last theorem, 47–48
Fermat's little theorem, 27–28, 178–179
Ferrari, Lodovico, 101, 102
Ferro, Scipiono del, 101
Ferrucci, David, 224
Feynman, Richard, 112–113, 310, 314–315
Fibonacci, 69–75, 81, 167
Fibonacci cubes, 74
Fibonacci heap, 74, 263

Fibonacci's numbers (or Fibonacci sequence), 69, 72–75
Fingers for counting, 2, 10
Finite automaton (FA), 207–211
Flint, Charles R., 95
Floating-point arithmetic, 271
Flowers, Tommy, 146, 151
Floyd. Robert, 263
Floyd's algorithm, 263
Ford, Lester R., 263
Formal languages, 211–214
Forrester, Joy, 158
Fortran, 231, 232–233
Fortran List Processing Language (FLPL), 234
Foster, Ian, 290
Fourier, Joseph, 102
Fractions
   ancient mathematics, 15–17
   decimal fractions, 77–78
   decimal notation, 18
   multiplication algorithm, 71
   unit fractions, 15
Freddy and Freddy II, 223
Fredman, Michael L., 263
Functional programming, 235, 260
Functions, 130
   recursive, 133
   total, 130–131
Fundamental theorem of algebra, 112
Fundamental theorem of arithmetic, 25
Fuzzy logic, 222, 223, 228

G

Galileo, 40
Galois, Evariste, 102, 103–105
Galois field, 198
Game theory, 170
Garey, Michael R., 252
Gauss, Johann Carl Friedrich, 29–30, 98, 111, 165, 176, 250
Gaussian elimination method, 98–99, 165–166
Gelernter, Herbert, 222, 226
General Problem Solver, 222
Generative grammars, 210–214
Geometry Theorem, 222, 226

Gibbs, Josiah Willard, 186
Gibbs entropy, 186
Ginsburg, Seymour, 210
Girard, Albert, 74
Go, 227
Gödel, Kurt, 118, 131, 135, 303
    incompleteness theorem, 129–130
Golay, Marcel J. E., 200
Golay code, 200, 201, 202
Goldbach, Christian, 28
Goldbach's conjecture, 28
Golden ratio, 73–74
Goldreich, Oded, 253
Goldstine, Herman, 170
Good, Irving John (Jack), 153
"Goto" statements, 263
Granville, Andrew, 33
Graph algorithms, 260–264
Graph isomorphism problem (GIP), 253
Great Cipher of Louis XIV, 143
Greatest common divisor (GCD),
            Euclidean algorithm for finding,
            37–39, 71, 134, 245
Greedy algorithm, 262
Gregorian calendar, 13
Greibach, Sheila, 210
Grid computing, 290–291
Group theory, 104
Grover, Lou, 313
Grover's search algorithm, 313–314
Gunter, Edmund, 81

H

Hadamard gate (H gate), 312, 313
Hamming, Richard Wesley, 195
Hamming codes, 196–198, 199, 201
Hamming distance, 195
Hansch, Michael Gottlieb, 87
Hardy, G. H., 181
Hartley, R. V. L., 183
Hartmanis, Juris, 249
Heath Robinson, 146, 152, 220
Hebb, Donald O., 221
Heisenberg, Werner, 309
Hellman, Martin, 298–299, 305
Hermite, Charles, 103, 109
Herodotus, 51–52

Heron of Alexandria, 110, 205
Hieroglyphic symbols, 7
Hilbert, David, 48, 109, 118, 129, 131, 181
Hilbert's problems, 48
Hindu-Arabic numerals, 8, 9, 13
    Brahmagupta and, 100
    Fibonacci and, 69, 70
    zero, 9, 12
Hippasus, 19, 161
Hitachi, 223
Hoare, Charles Antony Richard, 256–257,
    265
Hocquenghem, A., 201
Hollerith, Herman, 90, 93–95
Homophonic substitution ciphers, 141–143
Horsman, Clare, 316
Horswill, Ian, 223
Huffman, David Albert, 188, 207
Huffman coding, 188–189
Huskey, Harry, 158
Hypertext, 287–289
Hypertext Markup Language (HTML), 289
Hypertext Transfer Protocol (HTTP), 289

I

IBM 701, 156, 226, 232
IBM 7030, 271–272
IBM 704, 232–233, 271
IBM Sequoia, 274–275
ILLIAC IV project, 273
Imaginary numbers, 110–111
Imperative programming, 233–234
Impossibility theorem, 103
Incompleteness theorem, 129–130
Indian (Asian) numerals, See Hindu-Arabic
    numerals
Inference AI, 229
Infinite sets, 116–117
Information Processing Language, 221, 234
Information theory
    error correcting codes, 194, See also
            Error detecting and correcting
            codes
    Huffman coding, 188–189
    Morse code, 189–190

Shannon and, 126, 182, 183–186, 194, 225–226, *See also* Shannon, Claude Elwood
Shannon-Fano coding, 186–188
Wiener and cybernetics, 181–183
Institute for Advanced Study (IAS) computer, 156
Intergalactic Computer Network, 285
International Business Machines (IBM), 95
Internet
    origins, 285–287
    security issues, 297–299, *See also* Public-key cryptography
    World Wide Web, 287–289
Internet Protocol (IP), 286
Interpolation, 171
I/O automaton, 279
Irrational numbers, 19–21, 108–109, 161–162
    incommensurable numbers, 39
Ishango bone, 3
Ishii, Hiroshi, 291–292
Iterative techniques, 165–167
Iverson, Kenneth E., 259–260

J

Jacobi iteration, 167
Jacquard, Joseph Marie, 89–90
Jacquard loom, 89–91, 93, 219
Japanese abacus, 65–66
Japanese automata, 206
Japan's Newton, 169
Java, 241
Jevons, William Stanley, 122
Jiushao, Qin, 77
Johnson, David S., 252
Johnson, William Ernest, 122
Jones, William, 162

K

Kahn, Robert Elliot, 286
Kama Sutra, 55–56, 137
Kantorovich, Leonid, 169
Karatsuba, A., 248
Karnaugh, Maurice, 127
Karnaugh map, 127
Karp, Richard, 252, 253, 266

Kasiski, Friedrich Wilhelm, 141
Kasiski test, 141
Kauffman, Louis, 183
Kayal, Neeraj, 302–303
K computer, 274–275
Keating, Jon, 23
Kemeny, John G., 238
Kepler, Johannes, 40, 78, 81, 86
Kesselman, Carl, 290
Key distribution problem, 297
Kilburn, Tom, 157
Kiyasu, Zen-ichi, 201
Kleene, Stephen, 135, 209, 245
Kleene star, 210
Kleinrock, Leonard, 284, 290
Knuth, D. E., 38, 74, 245, 257–258
Kobliz, Neal, 303
Kolmogorov, Andrey Nikolaevich, 182
Kondo, Shigero, 164
Königsberg, Johannes Müller von, 47
Königsberg bridge problem, 260–262
Kuroda, S. Y., 214
Kurtz, Thomas E., 238

L

Lagrange, Joseph Louis, 102
Lambert, Johann, 162
Lamport, Leslie, 280
Landau, Edmund, 181
Landweber, Lawrence, 286–287
LARC, 272
Las Vegas algorithms, 266, 268
Law of cosine, 78
Lebombo stick, 3
Legendre, Adrien-Marie, 29, 103, 108
Leibniz, Gottfried Wilhelm von, 88–89, 111, 151, 167, 219
Levin, Leonid, 252
Liber Abaci (Fibonacci), 69–72, 81, 167, 177–178
Licklider, Joseph Carl Robnett, 284–285
Lincoln, Abraham, 40
Lindsay, Robert, 222
Linear algebra notation, 310
Linear bounded automata (LBAs), 211
Linear codes, 198–202
Linear equations, 73, 98–99

Linguistic analysis and cryptanalysis, 138–139
Linked lists, 258–260
Liouville, Joseph, 108
LISP, 222, 223, 226, 231, 234–235, 260
List ranking, 277–278
Lists, 258–260
Livermore Automatic Research Computer (LARC), 272
Llull, Ramon, 219
Logarithms, 79–82, 91
Logical AI, 228
Logical OR and AND, 121–122
Logical programming, 239
Logic Theorist, 221
Long Count Calendar, 13
Lorenz cipher machines, 145–146, 151–152
Lovelace, Ada, 90, 92–93
Lynch, Nancy A., 279

M

Machine code, 231–232
Mamdani, Abe, 222
Manaechmus, 35
Manchester Baby, 157
Manchester Mark I computer, 157–158
Marshack, Alexander, 2–3
Marten's function, 31
Matiyasevich, Uri Vladimirovich, 48
Mauchly, John, 153
Mayans, 13
McCarthy, John, 218, 221, 222, 226, 260
McCluskey, Edward J., 127
McCulloch, Warren S., 207
McDermott, John, 223
McEliece, Robert, 303
McEliece cryptosystem, 303
Mealy, George H., 207
Mechanical computers, 91–92
Memex, 287
Menninger, Karl, 8
Mesoamerican Long Count Calendar, 13
Meziriac, Claude-Gospar Bachet de, 47
Michie, Donald, 153, 218, 223
Miller, Gary Lee, 268
Miller, Raymond E., 302
Miller, Victor S., 303

Miller-Rabin primality test, 268, 302
Minimax theorem, 220, 226
Minsky, Marvin, 211, 221, 222, 227
Möbius, August Ferdinand, 30–31
Modular arithmetic
    Chinese remainder theorem, 175–178
    clock arithmetic, 173–175
    Fermat's little theorem, 178–179
Modularity theorem, 48
Moivre, A. de, 74
Monoalphabetic substitution cryptosystems, 54–55, 138–139
Monte Carlo algorithms, 266–268, 302
Moor, Edward F., 207
Moravec, Hans, 223
Morse, F. B., 189–190
Morse code, 189–190, 191t
Moscow Papyrus, 17
Mouri, Kanbei (or Shigeyoshi), 65–66
Muller, David E., 201
Multiplication
    algorithm in Liber Abaci, 71
    ancient methods, 15
    efficient algorithms for, 248
    Stepped Reckoner and, 88
    time complexity and efficiency of, 247–248
    use of logarithms, 79
Mutual exclusion problem, 280–281
MYCIN, 223, 228
Myhill, John R., 211

N

Napier, John, 78, 79–80
Napier bones, 81, 85
National Bureau of Standards (NBS), 158, 295–296
National Institute of Standards and Technology (NIST), 295, 305
Natural language processing, 228
Natural logarithm, 79
Neanderthals, 3
Negative numbers, 100
    square root of, 110–111
Nelson, Theodor Holm, 287–288

Network model for parallel computing, 276
Networks, *See* Computer networks;
        Distributed computing
Neural network models, 207, 221, 227
Newell, Allen, 221, 222
Newman, Allen, 260
Newman, Max, 145, 146, 151–152, 157–158
Newton, Isaac, 40, 88, 167
Newton's method, 167–170
Nick's class (NC) problems, 278
Nicomachus, 26
Nondeterministic Turing machine, 251
Non-Euclidean geometry, 36
Nonsecret encryption, 305–306
NOR and NAND gates, 125
NP-complete problems, 251–253
Numbers, representation of, 7–8
    decimal systems, 9–10, 12–13
    fractions, 15–16
    Hindu-Arabic numerals, 8
    positional systems, 8–9
    Roman numerals, 10, 11*t*
Numerical methods, 161
    algebraic equations, 164–170
    ancient methods, 161–164
    computing pi, 162–164
    Gaussian elimination method, 165–166
    interpolation and extrapolation, 171
    iterative techniques, 165–167
    modern methods and problem domains,
        170–172
    Newton's method, 167–170
    optimization problems, 170–171
Nyquist, Harry, 183

O

Object-oriented programming, 238,
        239–240
Oettinger, Anthony, 210
Ofman, Yu, 248
One-way functions, 299
Optimization problems, 170–171
Ordinal numbers, 118–119
OR gates, 125
OR operations, 121–122
Oughtred, William, 81

P

Pacioli, Luca, 111
Packet switching networks, 283–284
Paleolithic evidence, 1–3
Palindromes, 211
Papadimitriou, Christos H., 278
Paper, 3
Papert, Seymour, 222, 227
Parallel computing, 271–272
    computational complexity, 278
    parallel algorithms, 275–279
    supercomputers, 273–275
Parallel merge sorting, 276–277
Parallel random access machine (PRAM)
        model, 275
Parity check codes, 193–194
Parity check matrix, 200
Partial differential equations, 171
Pascal (programming language), 88, 234,
        260
Pascal, Blaise, 87
Pascaline, 87–88, 219
Pask, Gordon, 182
Pattern recognition AI, 229
Pauli-gates, 312
Pebbles for counting/calculating, 4–5, 61
Peirce, Charles Sanders, 122
Pell, John, 47
Pell's equation, 47
Perfect numbers, 24, 26, 39
Perles, M., 214
pH, 82
Phi function, 179
Phrase structure languages, 213, 214
Pi (π), 78, 108, 109–110, 162
    computing, 162–164, 265
    π-day, 164
Pippenger, Nicholas, 278
Pitts, Walter, 207
Place-value, 8, 9
Planck, Max, 309
Plato, 19, 35
Poisson, Simeon, 102
Polly, 223
Polyalphabetic substitution ciphers, 139–141
Polybius checkerboard, 55–56
Polygonal numbers, 47

Polynomial time reduction, 252
Porta, Giovanni, 139
Positional number systems, 8–9
Post, Emil L., 245
Prange, E., 201
Prime numbers, 23
    defined, 23
    Euclid's *Elements* and, 24–26
    Fermat's little theorem, 27–28, 178–179
    history of, 23–28
    LISP, 223
    modern developments, 32–33
    perfect numbers, 24, 26
    primality tests, 267–268, 302–303
    prime number theorem, 29–31
    public-key encryption and, 28, 267–268,
        302–303
    relatively prime numbers, 27
    Riemann hypothesis, 31
Procedural programming, 233–234
Proclus, 35
Programmable loom (Jacquard loom),
        89–91, 219
Programming languages, 231
    Abbreviated Computer Instructions, 157
    Ada, 93, 234
    ALGOL, 234, 236–237, 256
    algorithms and, 246, *See also*
        Computer algorithmic design
    APL, 259–260
    BASIC and Visual BASIC, 238
    C#, 242–243
    C++, 240–241
    COBOL, 237–238
    Fortran, 231, 232–233
    functional programming, 235, 260
    generative grammars, 214
    imperative and procedural
        programming, 233–234
    Information Processing Language,
        221, 234
    interpretative systems, 232
    Java, 241
    LISP, 222, 226, 231, 234–235, 260
    logical programming, 239
    machine code and assembly languages,
        231–232
    object-ordered programming, 239–240

Pascal, 88, 234, 260
PROLOG, 222, 239
Smalltalk, 240
structured programming, 88, 263
Project Xanadu, 288
PROLOG, 222, 239
Prony, Gaspard de, 91
Pseudocodes, 231–232
Public-key cryptography
    digital signatures, 303–305
    nonsecret encryption, 305–306
    origin and basic concepts, 297–299
    primality tests, 267–268
    prime numbers and, 28
    RSA, 38, 300–303
Punched cards, 90–91, 94–95
Pushdown automata (PDAs), 210–211
Pythagoras, 19, 35, 108, 161
Pythagoras's constant, 20, 162
Pythagoreans, 24
Pythagorean theorem, 19, 36
Pythagorean triples, 47

Q

Quadratic equations, 73, 97, 99–100
Quadratic reciprocity theorem, 30
Quantum computing, 309–311
    algorithms, 312–315
    difficulties and limits of, 315–316
    logic and gates, 312
    notation, 310–311
Quantum entanglement, 311
Quartic and quintic equations, 101–105
Qubit, 310–311
Quicksort, 256–257
    randomized, 265, 266, 268
Quine, Willard van Orman, 127
Quine-McCluskey algorithm, 127

R

Rabin, Michael Oser, 266, 302
RAND Corporation, 284
Random access memory (RAM), 275
Random error checking codes, 201–202
Randomized algorithms, 265–268
Raphson, Joseph, 169

Rational numbers, 17–19
   commensurable numbers, 39
Ray-Chaudhuri, D. K., 201
Real numbers, 107–110
Rechen Uhr ("calculating clock"), 86–87
Recursive algorithms, 257
Recursive functions, 133
Recursively enumerable sets, 213, 214
Reed, Irving S., 201
Reed-Muller codes, 201
Regula falsi, 167
Regular grammars, 214
Regular polyhedrons, 39
Regular sets, 209
Relatively prime numbers, 27
Representation AI, 229
Rhind, Alexander H., 16
Rhind Papyrus, 16, 24
Riemann, Georg Friedrich Bernhard, 31
Riemann hypothesis, 31
Right-linear grammars, 214
Rivest, Ron Linn, 300
Roberts, Laurence G., 285
Robotics laws, 221
Robots, 220, 222, 223–224
Rochester, Nathaniel, 221, 226
Roman hand abacus, 62–63
Roman numerals, 10, 11*t*, 13
Rosenblueth, Arturo, 181
Rossignol, Antoine, 143
Roussel, Philippe, 222, 239
Royce, Josia, 181
RP complexity class, 268
RSA cryptography, 38, 300–303
   digital signatures, 303–305
   nonsecret encryption, 305–306
Russell, Bertrand, 40, 181
Russell's paradox, 119

S

Salamis Tablet, 61
Samuel, Arthur, 222, 226
Sand abacus, 9, 59
Satisfiability problem (**SAT**), 252
Saxena, Nitin, 302–303
Scheme, 235
Scherbius, Arthur, 143

Schickard, Wilhelm, 86–87
Schmandt-Besserat, Denise, 5
Schnorr, C. P., 305
Schönhage-Strassen algorithm, 248
Schreyer, Helmut, 151
Schrödinger, Erwin, 309
Schützenberger, Marcel Paul, 210
SCON, 223
Scott, Dana, 266
Scytale, 53–54
SEAC, 158
Search techniques, 228, 257
   Grover's search algorithm, 313–314
Secret keys, 53
Secret writing, in ancient civilizations,
    52, 137, *See also* Cryptography
   cryptography, 52–56
   Kama Sutra and, 56
   steganography, 51–52
Sets
   algebra of, 122–123
   cardinality concepts and, 115–120
   regular, 209
Sexagesimal (base-60) number system,
    7, 8, 10, 60
Shakey, 222
Shamir, Adi, 300
Shamir, E., 214
Shannon, Claude Elwood, 126–127, 151,
    182, 183–186, 194, 195, 220, 221,
    225–226
Shannon entropy, 185
Shannon-Fano coding, 186–188
Shared memory model, 275, 279–280
Shaw, J. C., 221, 222
Shift cipher, 54–55, 137
Shor, Peter W., 315
Shor's factoring algorithm, 314–315
Shortest-path problems, 264
Shortliffe, Edward, 223
Show, Cliff, 260
Sierpinski, Waclaw Franciszek, 32
Sierpinski number, 32
Sieve of Eratosthenes, 26–27, 302
Simon, Herbert, 221, 222, 223, 260
Simpson, Thomas, 169, 171
Simpson's rule, 171

Single instruction, multiple data (SIMD), 273
Sipser, Michael, 253
Sketchpad, 285
Slepian, David S., 201
Slide rule, 81
Slotnick, Daniel L., 273
Smalltalk, 240
Smith, Adam, 91
Society of mind, 227
Solomon, 273
Solovay, Robert M., 268, 302
Solovay-Strassen primality test, 268, 302
Soroban, 66
Sorting algorithms, 255–258
    parallel merge sorting, 276–277
    randomized, 265, 267, 268
Space applications, 202, 283
Speech recognition systems, 223, 227
Speedcoding system, 232
Sputnik I, 283
Square root of 2, 20–21, 108, 161
Square root of negative number, 110–111
"Squaring the circle," 109
Standardization of cryptosystems, 295–296
Standards Eastern Automatic Computer (SEAC), 158
Standards Western Automatic computer (SWAC), 158
Stanford Cart, 223
State transition diagram, 208–209
Steam engines, 205
Stearns, Richard, 249
Steganography, 51–52
Steinmetz, Charles Proteus, 113
Stepped Reckoner, 88–89
Stevin, Simon, 78
Stibitz, George, 150
Stifel, Michael, 79
Stone, Marshall Harvey, 123
Strassen, Volker, 248, 268, 302
Straus, Ernst G., 47
STRETCH, 271–272
Structured programming, 88, 263
Suanpan, 64–65
Substitution cryptosystems, 53–55, 138–139
    homophonic, 141–143
    polyalphabetic, 139–141

Suetonius, 55
Sumerians, 5, 7, 10, 59–60
Sunzi Suanjing, 176–177
Supercomputers, 272, 273–275
Sutherland, Ivan, 285
SWAC, 158
Swift, Jonathan, 219
Switching theory, 127
Sylvester, James Joseph, 179
Symmetric-key cryptography, 296–297
Systems of linear equations, 98, 165–166
Systems of nonlinear equations, 169

T

Tabulator, 95
Takakazu-Kowa, Seki, 169
Tally marks and sticks, 2–4
Tangible Media Group, 291–292
Taniyama-Shimura-Weil conjecture, 48
Tarjan, Robert E., 263
Tartaglia, Nicolo Fontana, 101
Taube, Mortimer, 222
Taylor, Robert, 285
Teaetetus, 39
Thales, 35
Theaetetus, 35
Theon of Alexandria, 40
THEORYNET, 286–287
Tianhe-1A, 274
Tietavainen, A., 201
Tilman, John, 146
Time hierarchy theorem, 249
Time-space trade-off, 250
Toffoli gate, 312
Torres, Leonardo, 220
Total functions, 130–131
Totally ordered set, 107
Transcendental numbers, 108, 109–110, 162
Transmission Control Protocol (TCP), 286
Transposition cryptography, 53–54
Trigonometric functions, 79–80, 91
Trithemius, Johannes, 52, 139
Tukey, John W., 184
Turing, Alan M., 131, 145, 150, 151, 155, 183, 221, 245, 249
    Church-Turing thesis, 134–136
    Manchester Mark I project, 158

Turing machines, 131–134, 135, 145, 206–211, 220, 224–225, 245, 246, 250–252
Turing test, 133, 221, 225
Tuttle, Mark R., 279

U

Ubiquitous computing, 291–292
Ulam, Stanslaw Marcin, 32, 266
Ullmer, Brygg, 292
Uniform Resource Locator (URL), 289
Unit fractions, 15
UNIVAC, 154
Universal Automatic Computer (UNIVAC), 155
Universal Document Identifier (UDI), 289
Universal Turing machines, 134, 225

V

Vacuum cleaning robot, 224
Vail, Alfred, 190
Van der Waerden, B. L., 38
Van Lint, J. H., 201
Van Meter, R., 316
Vatsyayana, Mallanga, 56
VB.NET, 238, 243
Vector processing, 273
Vector representation, 111–112
Venn, John, 124
Venn diagrams, 124
Vignere, Blaise de, 139–140
Vignere cipher, 139–141
Virtual machines, 218, *See also* Artificial intelligence
Vision programs, 228
Visual BASIC, 238, 242
Von Lindemann, Ferdinand, 109
Von Neumann, John, 154, 155, 170, 220, 226

Von Neumann architecture, 155–156, 170, 220, 275

W

Wallis, John, 167
Watson, 224
Watson, Thomas J., 95
Weaver, Warren, 183
Weierstrass, Karl, 108–109
Weiser, Mark, 291–292
Wessel, Caspar, 111
Whirlwind I, 158
Wiener, Norbert, 181–182, 184
Wiener-Kolmogorov filter theory, 182
Wiles, Andrew John, 48
Wilkes, Maurice, 158
William of Ockham, 124
Williams, Frederic Calland, 157
Williamson, Melcolm John, 306
Wirth, Niklaus, 88
Womersley, John Ronald, 156–157
World Wide Web (WWW), 287–289

Y

Y-cruncher, 164
Yee, Alexander, 164
Yngve, Victor H., 260
Yoshida, Mitsuyoshi, 66

Z

Z3 computer, 150–151
Zadeh, Lotif, 228
Zairja, 219
Zero, 9, 10–13, 70, 100
ZPP complexity class, 268
Zuse, Konrad, 150

Milton Keynes UK
Ingram Content Group UK Ltd.
UKHW031142141024
449569UK00024B/1127